ELEMENTS OF GEOMETRY,

CONTAINING

THE FIRST SIX BOOKS OF EUCLID.

ELEMENTS OF GEOMETRY,

CONTAINING

THE FIRST SIX BOOKS OF EUCLID.

BY

PHILLIPS EUCLID

GYAN

Published by

Gyan Publishing House
5, Ansari Road
Daryaganj, New Delhi-110002
Phone: 011-47034999, 9811692060
E-mail: books@gyanbooks.com

Distribution Network
gyanbooks.com
India, USA, Canada, UK, Australia

ISBN: 978-93-6280-475-4 (HB)
First Published, 1826

2nd Impression 2024

Printed at: Gyan Press, Delhi.

ELEMENTS OF GEOMETRY, CONTAINING THE FIRST SIX BOOKS OF EUCLID
Author: PHILLIPS EUCLID

ELEMENTS OF GEOMETRY,

CONTAINING

THE FIRST SIX BOOKS OF EUCLID,

Translated into English, from the Edition of Peyrard.

TO WHICH ARE ADDED,

ALGEBRAIC DEMONSTRATIONS AND DEDUCTIONS;

WITH

NOTES, CRITICAL AND EXPLANATORY.

BY GEORGE PHILLIPS,

QUEEN'S COLLEGE, CAMBRIDGE.

LONDON:

PRINTED FOR BALDWIN, CRADOCK, AND JOY.

1826.

TO

OLINTHUS GREGORY, Esq. LL. D.

HONORARY MEMBER OF THE ACADEMY OF SCIENCES OF DIJON,

OF THE LITERARY AND PHILOSOPHICAL SOCIETY OF NEW YORK,

AND OF THE HISTORICAL SOCIETY OF THE SAME PLACE;

ONE OF THE COUNCIL OF THE ASTRONOMICAL SOCIETY OF LONDON, &c. &c.;

AND PROFESSOR OF MATHEMATICS IN THE ROYAL MILITARY ACADEMY, WOOLWICH;

AS A PLEDGE OF ESTEEM FOR PAST KINDNESS

THIS EDITION OF

EUCLID'S ELEMENTS

IS DEDICATED BY

HIS MOST OBLIGED

AND OBEDIENT SERVANT,

THE EDITOR.

PREFACE.

THE object of this edition of the ELEMENTS of EUCLID is to present an accurate translation of the twelve books, upon such a plan, and with such illustration, as may facilitate the advancement of the student.

In the accomplishment of the latter object, the Editor has inserted a number of deductions at the end of their respective propositions, by way of exercises in developing the powers of genius and inquiry; and he hopes that, as the student performs these in his progress through the work, they will serve to render the subject of Elementary Geometry more familiar to his mind. In the selection, the Editor has taken some from Cresswell, Bland, &c. and added others of his own; and he trusts it is made so as to meet the approbation of the public in general. There are also algebraic demonstrations annexed to the second and fifth books; for in these the Editor believes that analysis is generally employed as well in the Universities as in military and naval institutions; and, considering the facility which it affords, and the simplicity of its operations, no wonder that such a system should be universally adopted.

When the Editor first thought of undertaking the work, he purposed to make his translation from the Oxford copy, edited by Dr. David Gregory in 1703; but, after consulting the edition recently published at Paris under the superintendance of Peyrard, and reading the *lectiones variantes* of that work, the Editor fully resolved to make his translation from it: first, because it came out under the strongest recommendations of the best mathematicians on the Continent, such as Lagrange, Legendre, &c.; and, secondly, because the learned M. Peyrard himself bestowed the greatest labour in examining and collating all the existing MSS. and oldest editions.

The Editor has bestowed the greatest care in the execution of his undertaking; he has availed himself of the assistance of several eminent mathematicians; and he trusts that the public, in reviewing his labours, will, after an impartial criticism, be enabled to bestow upon him some commendation, the only reward which he can hope to receive.

INTRODUCTION.

THE term Geometry is derived from two Greek words, which literally signify the *art of measuring the earth:* it does not, however, so much imply the ascertaining the measure of the whole globe as that of certain parts of its surface; and hence we are informed by historians that the finding of the dimensions of lands, and other plane figures, with some of the most simple and obvious methods of determining their contents and relative proportions, were the first uses made of this science by the ancients. It has, however, since been extended to numberless other speculations; insomuch that, together with *analysis,* Geometry forms the principal foundation of all the mathematics.

Like many other arts and sciences, the origin of Geometry is involved in considerable obscurity, some authors fixing it at one period, and others at another. Most, however, assign Egypt for its birth-place, and that the annual inundations of the Nile first excited attention to this science among the inhabitants of that nation; for the waters bearing away the boundaries of the land, in the lower and most fertile parts of the country, and laying waste their estates, the people were obliged to devise some method for ascertaining the property of each person after the waters had subsided, and to establish it upon principles that would serve as a guide to posterity.

Herodotus, however, the first who wrote history in prose, assigns its origin to a different cause. The following is the account he himself gives of what he learned respecting it at Thebes and Memphis: "I was told," says he, "that Sesostris divided the kingdom among all his subjects, and that he had given each an equal quantity of land, on condition of paying annually a proportionate tribute. If the portion allotted to any one were diminished by the river, he went to the king and told him of what had happened; the king then sent and ordered the land to be measured, that he might know what diminution it had undergone, and demand a tribute only in proportion to what remained. Here, I believe," adds Herodotus, "Geometry first took its rise, and that hence it was transmitted to the Greeks."

If we wished, says Bossut, to indulge in frivolous conjectures, we should carry back the origin of Geometry to the invention of the square and compasses, since it makes the greatest use of them in practice; but the same argument of their use, continues he, will lead us to suppose that these instruments were invented at the commencement of society. Indeed some such instrument must have been used in the earliest ages of the world, as the rudest operations of nature could not be effected without them. But if we fix the period when Geometry began to assume the character of a real science, we shall at once transport ourselves to Greece and the age of Thales.

This illustrious philosopher was born at Miletum about 640 years before Christ. After receiving the usual learning of his own country he travelled into Egypt, where he became eminent in Astronomy, Geometry, Philosophy, &c. Whatever instructions he might receive from them in many branches of knowledge, it does not seem probable that he obtained much information from them in Geometry, as all writers agree that he was the first who measured the height of the pyramids by the extent of their shadows. It is said that he also applied the circumference of the circle in measuring angles. There can be little doubt that he made many other discoveries, which have not been directly handed down to us, but which might have been inserted in elementary books, and ranked among discoveries of later times; from him also Astronomy made a very considerable advance; and he is generally reputed to be the father of the Greek philosophy, being the first that made any researches into natural knowledge.

From Thales we pass on to Pythagoras, a philosopher no less distinguished than the former for the variety and extent of his discoveries; among the most eminent of these, at least in Geometry, which he made, may be mentioned, that the square of the hypothenuse of a right angled triangle is equal to the sum of the squares of the other two sides (see note to the 47th proposition of the 1st book); he is also said to be the inventor of the 32d proposition of the same book, viz. that the three angles of any triangle are together equal to two right angles; as likewise to have shown that only three polygons, or regular plane figures, can fill up the space about a point; viz. the equilateral triangle, the square, and the hexagon; these, however, when compared with his other inventions, will appear but trifles: in Astronomy he is reported to have maintained the true system of the world, which places the sun at the centre, and makes the planets to revolve round him; and from him it is called the Pythagorean system, which was revived by Copernicus. Whether we consider the variety of his discoveries, or the extent of his attainments; whether we reflect upon his inventive genius, which distinguishes him in all his pursuits, or upon that amazing assiduity so conspicuous amongst the whole race of Grecian philosophers, few men will be found to possess a greater claim to the honour of posterity than Pythagoras. As a mathematician. he was decidedly the first of his time. As a philosopher, we find him delivering many excellent things concerning God and the human soul, and a great variety of precepts relating to the conduct of life both political and civil.

Although it is doubtful whether Geometry at this time had been founded into a regular system, yet, from what has been said, it appears that it must certainly have made considerable advances, and that many of its detached parts were known; for not more than a century had elapsed, from the age of Pythagoras, when Zenodorus, a man of great parts, arose, and whose writings are the first amongst the ancients, which have survived the wreck of time, a geometrical tract of his having been preserved by Theon in his Commentary upon Ptolemy's Almagest, wherein he has shown the falsity of the opinions then entertained that figures, with equal peripheries, have equal areas: a problem not easy of solution, and shows that Geometry must have then made a great progress. The ingenious theory of the five regular bodies originated also about the same time in the Pythagorean school.

Next in order comes Hippocrates, a man possessing a very brilliant genius, and who rendered Geometry essential services by his diligence and assiduity. Among his discoveries the quadrature of the celebrated lunulæ of the circles, which bear his name, may be placed foremost in the list. Having described three semicircles on the three sides of a right angled triangle considered as diameters, the one on the hypothenuse being in the same direction as the others, he found that the sum of the areas of the two equal lines comprised between the two quadrants to the hypothenuse, and the semicircles answering to the other two sides, was equal to the area of the triangle. He also wrote Elements of Geometry; which, from the account given by Proclus, were much esteemed in his time; although, having been superseded by the Elements of Euclid, those of Hippocrates were consigned to oblivion. He also appears with honour among the list of Geometers, who attempted to solve the celebrated problem of the duplication of the cube, which at that period began to be pursued with ardour. The circumstance of this problem is well known; its solution at first sight appeared easy; but the mistake was soon perceived, and all the geometricians of Greece were baffled in attempting to solve it. Notwithstanding the failure of this, Geometry still continued to advance, and was cultivated with great care and attention by Plato (B. C. 300); although we have no work of his written upon this subject, yet we learn from other writers, and indeed from many passages in his works, that he was well acquainted with its different branches, and had even enriched it with many of his discoveries. Indeed so profound a veneration did he entertain for the science here spoken of, that he made it the principal object of instruction among his scholars. He had written over the door of his academy, "Let no one enter here who is ignorant of Geometry."

The problem before-mentioned, viz. the duplication of the cube, particularly engaged his attention; and although he was unable to resolve it by a method purely geometrical, or at least as considered by the ancients, that is, by the rule and compasses only, yet he invented an ingenious, though mechanical solution, by means of an instrument consisting of two rules, one of them moved in the grooves of two arms at right angles with the other, so as always to continue parallel with it. But though Plato was unfortunate in his attempts to double the cube, yet we find him more successful in another speculation of a kind entirely new. Before his time the circle was the only curve admitted into geometry, Plato, however, discovered the conic sections, or those curves which are found on the surface of a cone by a plane cutting it in different directions; and by attentively examining the generation of those curves, he discovered some of their most remarkable properties, which being made the continual study of his scholars and successors, it at length became a distinct science from the common elements, and received the appellation of the *higher* or *sublime* geometry.

The important addition of conic sections to the mathematical sciences being, as before observed, particularly cultivated by the geometers of that time; Aristeus, a friend and disciple of Plato, composed five books on that subject, which are spoken of with great eulogies by the ancients; but either from the despotism of ignorant barbarians, or from the ravages of time, they have unfortunately not reached us, and nothing more is known of them than the little that is mentioned by Pappus, in his Mathematical Collections. Of Menechmus, we have two learned applications of the same theory to the problem of the duplication of the cube; and from the result of his labours, it appears that if we possessed the means of describing conic sections by one continued motion, in as simple a way as we trace a circle with the compasses, the solutions of Menechmus would have all the advantage of geometrical construction in the sense which the ancients applied to the term. But at present no instruments have been made that will describe the conic sections in this manner.

I cannot, however, pass over the problem of the trisection of an angle, which is of the same kind with that of doubling the cube, both of which were equally agitated in the school of Plato; and although a solution was not to be attained by means of the rule and compasses only, yet it was reduced to a very neat and simple proposition. This consists in drawing a right line from a given point to the semi-periphery of a circle, which line shall cut this periphery, and the prolongation of the diameter that forms its base, so that the part of the line comprised between the two points of intersection shall be equal to the radius; a result that gives rise to many simple constructions, two of which may be seen in Bonnycastle's Elements of Geometry, page 282. Most of the ancients, however, were so possessed with the hope of solving these two problems with the rule and compasses only, that they could not be persuaded to give it up; they made many fruitless endeavours; and this anxiety raged like an epidemic disease, which has been transmitted from age to age down to the present day; but after they have baffled the attempts of such illustrious characters as Archimedes among the ancients, and Newton and Maclaurin among the moderns, it would surely argue a want of discretion in a young mathematician to waste his time in such ill-fated speculations.

It may not be improper to mention the celebrated Aristotle, the successor of Plato, and preceptor to Alexander the Great; but though, in other respects, he may be regarded as one of the greatest men of his or indeed of any other time, yet we are not told that he made any improvements in the mathematical sciences. After attending the lectures of Plato, he opened a school himself in the Lyceum of Macedonia, which was assigned him by the magistrates, and was the founder of the sect called the Peripatetics. Among the number of his disciples were Theophrastus and Eudemus, who particularly applied themselves to the study of the mathematics. The former wrote a History of the Mathematics in eleven books, from their origin to his own time, four of which treated on Geometry, six on Astronomy, and one on Arithmetic. The latter also wrote a work of a similar kind, consisting of six books, on the History of Geometry, and another of the same number of books on that of Astronomy. But these, which would have been so useful to the modern scientific inquirer, which would have assisted him so much in his researches after the precise origin of the various mathematical sciences, and their progress in those times, have not been transmitted to the present age.

Notwithstanding the ancients were not successful in the object they sought to attain, yet Geometry received additional splendor from the researches they were continually making; new theories were introduced, and some ingenious instruments for solving the two problems in question, so as to approximate the truth near enough for practical purposes; most of these methods are now lost, but those of four eminent geometricians, viz. Dinostratus, Nicomedes, Pappus, and Diocles, deserve particular praise for their merit; but the reader must excuse my not entering into an explanation, or exhibiting to him a view of their several plans, as such would swell this introduction much beyond the limits which I intend it should occupy.

Next after the period of Plato, and his disciples here mentioned (passing over Euclid for the present), may be reckoned Archimedes of Syracuse, who was born about 280 years before Christ. In his youth, he devoted himself to the study of Geometry; and in his maturer years, he travelled into Egypt, where the Greeks usually resorted in the pursuit of science. After an absence of several years, which he spent in the society of Conon and other eminent men, and during which time he gave very promising indications of his future fame, he returned into his own country, and then continued his studies with the greatest zeal and assiduity. Such, indeed, were the intenseness and ardour of his application to mathematical sciences, that he prosecuted his studies to the neglect both of food and sleep, and improved the minutest circumstance that occurred into an occasion of making very important and useful discoveries.

His active and comprehensive genius led him to the study of every branch of science then known; Geometry, Arithmetic, Optics, &c. equally engaged his attention, and alike experienced the powerful effects of his superior talents, talents which placed him with such distinguished lustre in the view of the world as to render him both the honour of his own age, and the admiration of posterity. He was, indeed, the prince of the ancient mathematicians, being to them what Newton is to the moderns, to whom, in his genius and character, he bears a very near resemblance. He was the first who squared a curvilineal space, excepting Hippocrates, on account of his *lunulæ*. He applied himself closely to the measuring of conic sections, as well as other figures. He determined the relations of spheres, spheroids, and conoids, to cylinders and cones, and the relations of parabolas to rectilineal planes, whose quadratures had long before been determined in geometry. He also proved that a circle is equal to a right angled triangle, whose base is equal to the circumference, and its altitude equal to the radius.

Being unable to determine the exact quadrature of the circle, for want of the rectification of its circumference, which all the methods he devised would not effect, he proceeded to assign a useful approximation to it: this he effected by the numeral calculation of the perimeters of the inscribed and circumscribed polygons; from which calculation it appears, that the perimeter of the circumscribed regular polygon of 192 sides is to its diameter in a less ratio than $3\frac{1}{7}$ to 1, and that the perimeter of the inscribed polygon of 96 sides is to the diameter in a greater ratio than that of $3\frac{10}{71}$ to 1: therefore the ratio of the circumference to its diameter must be between these two ratios. But that which has rendered him most famous in the eyes of posterity is the fabrication of such admirable engines for the defence of Syracuse when besieged by the Roman consul Marcellus, showering upon the enemy sometimes long darts, and stones of vast weight, and in great quantities; at other times lifting their ships up into the air, that had come near the walls, and dashing them to pieces, by letting them fall down again; nor could they find their safety in removing out of the reach of his cranes and levers, for there he continued to fire them with the rays of the sun reflected from burning glasses.

However, Syracuse was at length taken; "what gave Marcellus the greatest concern," says Plutarch, "was the unhappy fate of Archimedes, who was at that time in the museum; so intent was his mind, as well as his eye, upon some geometrical figures, that he heard not the clashing of arms, nor the invasion of the city; in this state of abstraction, a soldier came suddenly upon him, and commanded him to follow him to Marcellus; but he refusing to stir till he had finished his problem so much enraged the soldier that he ran his sword through his body." Livy says, that Marcellus was so much grieved that he took care of his funeral, and made his name a protection and honour to those who could claim any relationship with him.

Archimedes was a lover of glory; not of that sordid ambition which inspires mediocrity, but of solid glory, which is due to a man who has enlarged the limits of science. He desired, when he was dying, that a sphere inscribed in a cylinder might be engraved on his tomb, to perpetuate the memory of his most brilliant discovery; the Sicilians, however, having their minds turned upon different objects than Geometry, forgot the man who was their chief honour in the eyes of posterity. Two hundred years after his death, Cicero being then quæstor in Sicily, gave, to use his own words, Archimedes a second time to light: unable to learn from the Sicilians the place of his interment, he sought for it by the symbol before mentioned, and six verses in Greek inscribed upon its base. After much fruitless research, it was at length discovered in a field near Syracuse over-grown with thorns; he showed it to the Sicilians, who blushed for their ignorance and ingratitude. Not more than fifty years had elapsed since the death of Archimedes, when Apollonius arose, who, if not equal to his illustrious predecessor, certainly ranks in the second place among the ancients, and who gave a great impulse to the mathematical sciences. He was born at Perga, in Pamphylia, whence he is called Apollonius Pergæus, to distinguish him from others of the same name. His contemporaries styled him *the Great Geometrician*, and posterity has confirmed this honourable title without detracting from the merit of Archimedes, to whom it assigns the first place.

Apollonius composed a great number of books, which were considered by the ancients as affording the most perfect examples of the higher geometry of that time; most of these are now lost, or exist only in fragments; we have, however, nearly the whole of his conic sections, which are alone sufficient to establish his fame, and to merit the title before-mentioned; this treatise consisted originally of eight books; the first four of which have been transmitted to us in the language in which they were written; and the following three had been preserved only in an Arabic translation made about the year 1250, and translated into Latin about the middle of the seventeenth century by Borelli; but to the great regret of all geometers, the eighth is entirely lost. A magnificent edition was published by Dr. Halley in folio, at Oxford, in 1710, together with the Lemmas of Pappus, and the Commentaries of Eutacius. The other writings of Apollonius, mentioned by Pappus, are,

1. The Section of a Ratio, or Proportional Section; two books.
2. The Section of a Space, in two books.
3. Determinate Section, in two books.
4. The Tangencies, in two books.
5. The Inclinations, in two books.
6. The Plane Lair, in two books.

Were I writing a minute history of mathematics, I might give an account of the geometricians, who flourished from the time of Archimedes to the destruction of the Alexandrian School; but as this introduction is intended only as a brief historical sketch of those ancient mathematicians, who successively improved and made discoveries in the sciences, the reader must not expect to find an enlarged history of an obscure individual, or a full relation of a trifling improvement.

It may not be improper, however, to name Conon and Dositheus, both very learned men, and both friends of Archimedes, Gemmius, a mathematician of Rhodes, who wrote a work entitled "Enarrationes Geometricæ," &c.

After these we may reckon Theodosius, who wrote a treatise on spherics, in which he examines the properties which circles formed by cutting a sphere in all directions have with respect to each other. From the time of this eminent man, we move on for three or four hundred years without meeting with one person who contributed anything to the advancement of the sciences. Theon, however, appeared about 380 years after Christ; and by his skill and perseverance in mathematics and philosophy, he obtained the honourable dignity of being appointed president of the famous Alexandrian School, where, by his erudition and conduct, he gained the greatest respect and reputation. His principal works, which have escaped the ravages of time, are his Scholia, or Notes on Euclid's Elements, and his Commentary on the First Eleven Books of Ptolemy's Almagest. They were published in Greek in the years 1633 and 1638. The

Scholia were published by Commandine in one of his Latin editions of Euclid. His Commentaries, however, on the Almagest have not yet been translated, except the first book.

One of his most celebrated pupils was his own daughter Hypatia, a very learned and beautiful lady, born at Alexandria about the end of the fourth century. Her father, perceiving her extraordinary genius, had her not only educated in all the ordinary accomplishments of her sex, but instructed in the most abstruse sciences. She made such great progress in philosophy, geometry, astronomy, and the mathematics in general, that she passed for the most learned person of her time. She published Commentaries on Apollonius's Conics, on Diophantus's Arithmetic, and other works. Whilst very young she was chosen to succeed her father in the same school, and to deliver instructions out of that chair, where Ammonius, Hierocles, and many other very learned men abounded, both at Alexandria and many other parts of the Roman empire. The pupils of this lovely and surprising female were not less eminent than they were numerous. Amongst whom was the much esteemed Synesius, afterward bishop of Ptolemais. But it was not Synesius only, and the disciples of the Alexandrian School, who admired Hypatia for her virtue and learning: never was woman more caressed by the public, and yet never had woman a more unspotted character. She was held as an oracle for her wisdom, for which she was consulted by the magistrates on all important cases. In short, when Nicephorus intended to pass the highest compliment on the princess Eudocia, he thought he could not do it better than by calling her another Hypatia. Whilst Hypatia thus reigned the most brilliant ornament of her sex in the annals of history, she was greatly admired by Orestes, the governor of that city, who, on account of her wisdom, often consulted her. This, together with an aversion which Cyril had against Orestes, proved the cause of her ruin. About 500 monks assembling, attacked the governor one day, and would have killed him had he not been rescued by the townsmen; and the respect which Orestes had for Hypatia, causing her to be traduced amongst the Christian multitude, they dragged her from her chair, tore her in pieces, and burnt her limbs. This shocking catastrophe was perpetrated in the Lent of the year 416. For a more particular account of this illustrious victim of fanaticism, see Bossut's History of the Mathematics, English edition, 8vo. 1803.

At length we come to Pappus, a consummate mathematician, who flourished towards the end of the fourth century, in the reign of Theodosius the Great: many of his works are lost, or lie in the hitherto unexplored recesses of public libraries; Suidas mentioned many of them, as also Vossius de Scientiis Mathematicis: amongst which, his Mathematical Collections, consisting of eight books, have transmitted his name with distinguished lustre to posterity. In them the author has assembled together a vast number of ancient works, most of which are now lost, and to these he has added several new, curious, and learned propositions of his own; six of the books were published by Commandine, in folio, in 1388, and a second edition of the same in 1660. In 1688, Dr. Wallis printed the last twelve propositions of the second book, in his Aristarchus Samius. In 1703, Dr. David Gregory gave part of the preface of the seventh book in the prolegomena to his Euclid. In 1706 Dr. Halley gave that preface entire in the beginning of his Apollonius. Among the researches of Pappus, may be named the problem of geometrical Loci, in which he advanced a very great way to its solution; but as a concise account of his collections will not convey to the reader any accurate idea of its contents, and as the prescribed limits of this introduction will not allow a prolix one, and since there is no English edition of his works, I beg to refer the reader to Hutton's Mathematical and Philosophical Dictionary, under the word Pappus; where he can receive the wished-for information.

We must not pass over Proclus, the head of the Platonic school, established at Athens A.D. 500. He rendered important services to the sciences, and showed great kindness to those who embraced their pursuit; he wrote a Commentary on the First Book of Euclid, which contains many curious observations respecting the history and metaphysics of geometry, as make us wish that he had extended his inquiries to the following books. His successor, Marinus, was the author of a Preface to Euclid's Data, which is generally prefixed to the head of that work. Hero also has given, in his Geodesia, the method of finding the area of a triangle by means of its three sides, without knowing the perpendicular.

Having thus far exhibited to the reader those votaries of science who flourished in the first period of her history, with their discoveries and improvements, let us turn back to Euclid, the celebrated author of the Elements of Geometry, which bear his name. The origin, and country of Euclid, are not fully known; although he is generally supposed to have lived in the time of Ptolemy Lagus, about 270 years before the Christian era; and it is also generally believed that he studied at Athens, under the disciples of Plato.

No book of science ever met with success equal to that of Euclid's Elements. They have been taught for several centuries in every mathematical school of eminence, and translated and commented upon in all languages.

With regard to the composition of the work, it is evident from the authority of Proclus, and other ancient writers, that many of the propositions contained in it were known at a very early period, and that elementary treatises on geometry had been composed by Hippocrates of Chios, Eudoxus, and many others, which doubtless rendered him considerable services in the formation of his Elements. As a proof of the superior excellence of this work, Euclid has deduced from a few first principles or axioms, a complete series of the most useful propositions in the science. His demonstrations are so very nervous and elegant, as not to be equalled by any geometrical writer, ancient or modern; and his method is such that nothing is taken as true unless demonstrated; and nothing is demonstrated, but from what went before. In consequence of this rigorous system of demonstration, it is reported that king Ptolemy, once asking Euclid whether there was no shorter way of arriving at geometry than by these his Elements, is said to have answered, *There is no other way or royal road to Geometry.*

Of the numberless editions of this valuable work, the following have met with the most considerable encouragement for their accuracy and superior excellence.

Campanus translated the whole fifteen books of the Elements into Latin, from the Arabic, in 1482.

Zambertus translated from Greek into Latin, the fifteen books and Data. This edition was edited at Paris in the year 1516; also at Basil in 1537, and 1546. The Data are only in the two last editions.

Candalla edited a Latin translation of the fifteen books in 1566.

Commandine, one of the best geometers of his age, translated into Latin the fifteen books from the Greek text of the Basil edition.

The Greek text of the Data of Euclid, with the Latin translation of Hardiæus, was edited in 1625.

A superb edition of all the works of Euclid, was edited in 1703, by Dr. David Gregory, in Greek and Latin, under the title of *Euclidis quæ supersunt omnia.*

Peyrard edited at Paris the fifteen books and Data in 1818, which is esteemed the best edition as to correctness and purity of text.

In English, we have Billingsley's, Barrow's, Keill's, Stone's, Simson's, &c. editions, which possess great merit, and which do honour to the talents of their respective editors.

Explanation of Characters used in the Work.

$+$ is the sign of Addition.

$-$ . . . . Subtraction.

$\times$ . . . . Multiplication.

$\div$ . . . . Division.

$=$ . . . . Equality.

$>$ signifies greater than.

$<$ . . . . less than.

$\therefore$ . . . . therefore.

EUCLID'S ELEMENTS.

BOOK I.

DEFINITIONS.

1. A POINT is that which has no magnitude, or is no part of any thing.
2. A line is length without breadth.
3. The extremities of a line are points.
4. A right line is that which lies evenly between its extreme points.
5. A superficies is that which has only length and breadth.
6. The extremities of a superficies are lines.
7. A plane superficies is that which lies evenly between its lines.
8. A plane angle is the mutual inclination of two lines to one another in the same plane, so touching each other as not both to lie in the same right line.
9. When the lines containing the said angle are right lines, it is called a rectilineal angle.
10. When a right line standing on another right line, makes the adjacent angles equal to one another, each of the equal angles is a right angle, and the right line standing on the other is called a perpendicular.

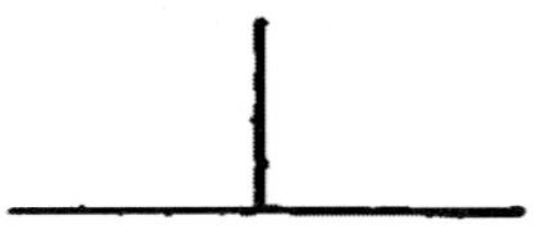

11. An obtuse angle is that which is greater than a right angle.

12. An acute angle is that which is less than a right angle.

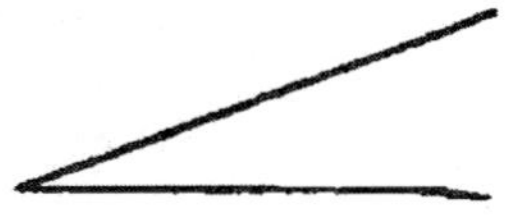

13. A term is the extremity of any thing.

14. A figure is that which is contained under one or more terms.

15. A circle is a plane figure contained by one line, which is called the circumference, to which all right lines drawn from one point within the figure are equal to one another.

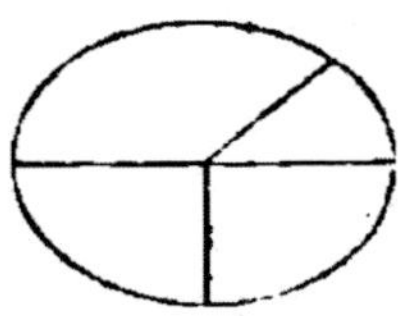

16. And this point is called the centre of the circle.

17. A diameter of a circle is a certain right line drawn through the centre, and terminated both ways by the circumference of the circle, and divides the circle into two equal parts.

18. A semicircle is the figure contained by the diameter, and the part of the circumference cut off by the diameter.*

19. Rectilineal figures are those which are contained by right lines.

20. Triangles are such as are contained by three right lines.

21. Quadrilateral, by four right lines.

22. Multilateral figures, or polygons, by more than four right lines.

23. Of trilateral figures, an equilateral triangle is that which has three equal sides.

24. An isosceles triangle is that which has only two equal sides.

25. A scalene triangle is that which has three unequal sides.

26. Of three sided figures, a right angled triangle is that which has a right angle.

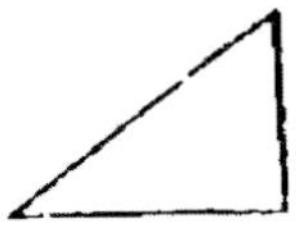

* The segment of a circle which is defined in this place, I have purposely omitted, as being of no use, until the third book, where the definition is repeated; instead of this Proclus has given in his Commentaries the following. *The centre of the semicircle is the same with that of the circle;* but as this is never used in the Elements, I have thought proper to reject it likewise.

27. An obtuse angled triangle is that which has an obtuse angle.

28. An acute angled triangle is that which has three acute angles.

29. Of four sided figures, a square is that which has its sides equal, and its angles right angles.

30. An oblong is that which has its angles right angles, but all its sides not equal.

31. A rhombus has its sides equal, but its angles not right angles.

32. A rhomboid has its opposite sides and angles equal to one another, but all its sides are not equal, nor its angles right angles.

33. All other four sided figures besides these are called trapeziums.

34. Parallel right lines are those which are in the same plane, and being infinitely produced either way, do not meet one another.*

POSTULATES.

1. Grant, that a right line may be drawn from any one point to any other point.
2. That a finite right line may be produced directly forwards.
3. That a circle may be described with any distance and from any centre.
4. That all right angles are equal to one another.†
5. That if a right line falling on two right lines make the interior angles at the same parts less than two right angles; these right lines being continually produced shall meet on that side where the angles are less than two right angles.
6. That two right lines cannot inclose a space.

* Newton in lemma 22, book 1, of his Principia, says, that parallels are such lines as tend to a point infinitely distant.

† For a demonstration of this, see Legendre's Geometry, proposition 1, book 1.

AXIOMS.

1. Things which are equal to the same are equal to one another.
2. If equals be added to equals, the wholes are equal.
3. If equals be taken from equals, the remainders are equal.
4. If equals be added to unequals, the wholes are unequal.
5. If equals be taken from unequals, the remainders are unequal.
6. Things which are double of the same, are equal to one another.
7. Things which are halves of the same, are equal to one another.
8. Things which mutually agree with one another, are equal to one another.
9. The whole is greater than its part.

PROPOSITION I.

PROBLEM.

Upon a given finite right line to describe an equilateral triangle

Let AB be the given finite right line; it is required upon AB to describe an equilateral triangle. From the centre A with the distance AB describe the circle BCD:[a] and again from the centre B, with the distance BA, describe the circle ACE, and from the point C in which the circles cut one another, draw the right lines[b] CA, CB, to *the points* A, B.

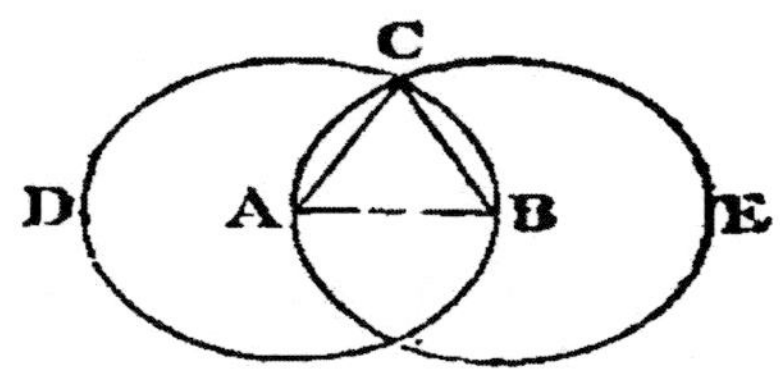

Therefore because A is the centre of the circle DBC, AC will be equal[c] to AB. Again, because B is the centre of the circle CAE, BC will be equal to BA: but it has been shown that CA is equal to AB: therefore CA, CB, are each of them equal to AB. And things which are equal to the same are equal to one another. Whence CA is equal to CB; wherefore the three, CA, AB, BC, are equal to one another; and, consequently, the triangle ABC is equilateral, and it is described upon the given finite right line AB. Q. E. F.

[a] Post. 3.
[b] Post. 1.
[c] Def. 15.

PROPOSITION II.

PROBLEM.

From a given point to draw a right line equal to a given right line.

Let A be the given point, and BC the given right line: it is required to draw from the point A a right line equal to the given right line BC. Draw the right line AC from the point A to C, and upon it describe the equilateral triangle DAC,[a] and produce[b] the right lines DA, DC, to E and F, and with centre C, and distance BC, describe the circle BGH.[c] Again, with centre D, and distance DG, describe the circle GKL: therefore because the point C

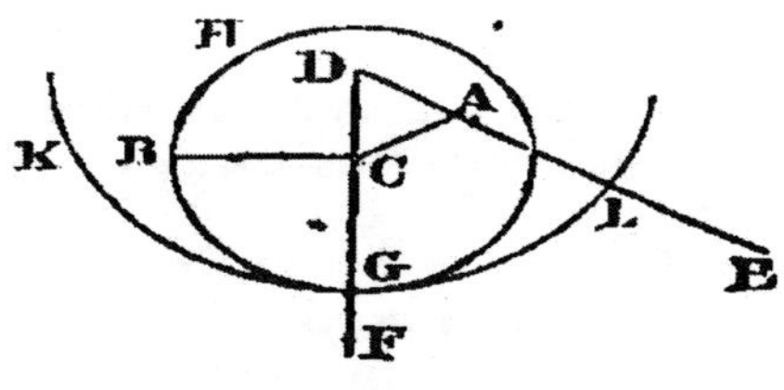

[a] 1. 1.
[b] Post. 2.
[c] Post. 3.

is the centre of the circle BGH, BC will be equal to
[d] Def. 15. CG.[d] Again, because D is the centre of the circles GKL, DL will be equal to DG and DA DC parts of them are equal: therefore the remainder AL is equal to the remainder CG. But it has been shown that BC is equal
[e] Ax. 3. to CG.[e] Wherefore each of them, AL, BC, is equal to CG. And things which are equal to the same thing are equal to one another. Whence AL is equal to BC. Therefore from a given point, AL has been drawn, &c. Q. E. F.*

PROPOSITION III.

Problem.

Two unequal right lines being given, to cut off from the greater a part equal to the less.

Let AB and C be two unequal given right lines of which AB is the greater: it is required to cut off from the greater, AB, a right line equal to C, the less. Draw from the point A a
[a] 2. 1. right line, AD, equal to C;[a] and from the centre, A, with the distance AD,
[b] Post. 3. describe the circle DEF.[b] And because A is the centre of the circle DEF, AD will be equal to AE. But AD is also equal to C. Therefore each of them, AE, C, will be equal
[c] Ax. 1. to AD. Wherefore AE is also equal to C.[c] Therefore two unequal right lines being given, &c.† Q. E. F.

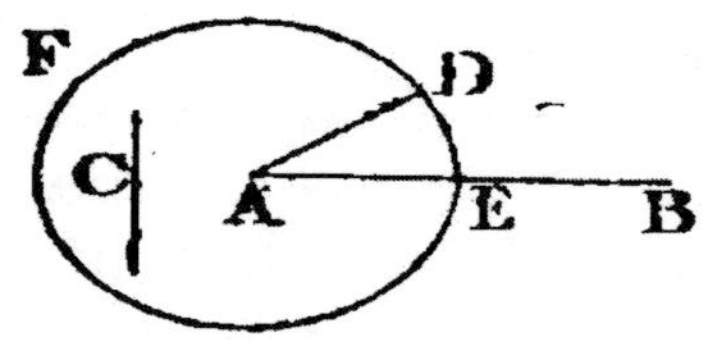

PROPOSITION IV.

Theorem.

If two triangles have two sides equal to two sides, each to each; and have also one angle equal to one angle, viz. that which is contained by the equal right lines: then shall the base of the one be equal to the base of the other; and

* This proposition may be divided into a variety of cases according to the different positions of the point A, although the construction and demonstration will, in every respect, be the same. Proclus remarks that some performed it by taking the line AL with a pair of compasses; but he by no means approved of the method, as those who thus reason, he says, *beg* in the very beginning.

† Some persons perform this proposition by taking the less line in the compasses, and with one leg in either extremity of the greater, cutting off with the other leg the part required: this, though correct in its operation, is certainly not geometrical, and would come rather under the class of postulates, than a demonstrable proposition.

one triangle equal to the other triangle; also the remaining angles of the one shall be equal to the remaining angles of the other, each to each, which are opposite to the equal sides.

Let there be two triangles, ABC, DEF, which have the two sides AB, AC, equal to the two sides DE, DF, each to each; namely, the side AB equal to the side DE, and the side AC equal to DF; also the angle BAC equal to the angle EDF. Then is the base BC equal to the base EF, and the triangle ABC equal to the triangle DEF; also the remaining angles equal to the remaining angles, each to each, to which the equal sides are opposite; namely, the angle ABC to the angle DEF; and the angle ACB to the angle DFE.

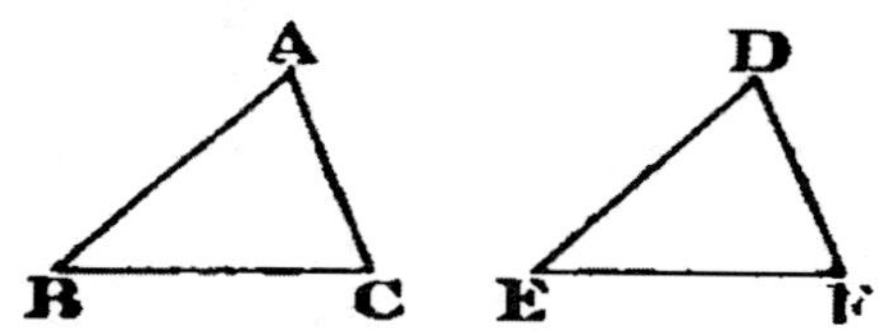

For if the triangle ABC be applied to the triangle DEF, and the point A be put upon the point D, and the right line AB upon the right line DE, then shall the point B coincide with the point E, because AB is equal to DE. But AB coinciding with DE; the right line AC shall also coincide with the right line DF, since the angle BAC is equal to the angle EDF. Wherefore C will also coincide with F: for the right line AC is equal to the right line DF; but the point B coincides with the point E. Therefore the base BC will also coincide with the base EF. Because if the point B coinciding with the point E, and C with F; the base BC does not coincide with the base EF; two right lines would inclose a space: which is impossible.[a] Whence the base BC coincides with the base EF, and also equal to it. Therefore the whole triangle ABC will coincide with the whole triangle DEF, and will be equal to it; also the remaining angles will coincide with the remaining angles, and equal to them,[b] viz., the angle ABC to the angle DEF, and the angle ACB to the angle DFE. Therefore, if two triangles have two sides of the one equal to two sides of the other, &c. Q. E. D.

[a] Ax. 10.

[b] Ax. 8.

PROPOSITION V.

Theorem.*

The angles which are at the base of isosceles triangles are equal to one another; and the equal right lines being produced, the angles under the base shall be equal to one another.

Let ABC be an isosceles triangle, having the side AB equal to the side AC, and produce the right lines AB, AC, directly forward to D, E. Then is the angle ABC equal to the angle ACB, and the angle CBD to the angle BCE. For take in the line BD any point F: and from the greater AE cut off AG equal to AF the less: also join FC, GB. Therefore, because AF is equal to AG; and AB to AC; the two FA, AC, are equal to the two GA, AB, each to each; and contain the common angle FAG. Therefore the base FC is equal[a] to the base GB, and the triangle AFC equal to the triangle AGB; also the remaining angles shall be equal to the remaining angles, each to each, viz., the angle ACF equal to the angle ABG; also the angle AFC to the angle AGB. And because the whole AF is equal to the whole AG; of which the parts AB, AC, are equal; the remaining part BF will also be equal to the remaining part CG. But it has been proved that FC is equal to GB. Therefore the two BF, FC, are equal to the two CG, GB, each to each; and the angle BFC equal to the angle CGB: also their base BC is common. Whence the triangle BFC is equal to the triangle CGB; and the remaining angles equal to the remaining angles, each to each, to which the equal sides are opposite. Therefore the angle FBC is equal to the angle GCB; and the angle BCF to the angle CBG. Wherefore, because the whole angle ABG has been proved to be equal to the

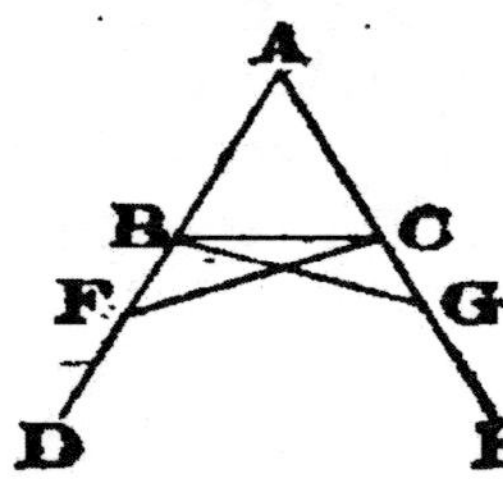

[a] 4. 1.

* This theorem was discovered by Thales, for he is first said to have perceived and proved, that the angles at the base of every isosceles triangle are equal, and, after the manner of the ancients, to have called them similar. The latter part of it is not at all necessary in demonstrating the former; and it is affirmed by some geometers, amongst whom is Scarborough, that it is not Euclid's, but added by some one else; however this may be, the angles opposite the equal sides may be demonstrated without proving the equality of the angles under the base, as is evident by the very elegant and concise demonstration of Pappus, and indeed by many others.

whole angle ACF, of which the angle CBG is equal the angle BCF; the remaining angle ABC[b] will be equal to the remaining angle ACB: and they are at the base of the triangle ABC. But it has also been proved, that the angle FBC is equal to the angle GCB, which are under the base. Therefore the angles which are at the base of isosceles triangles are equal to one another; and the equal right lines being produced, the angles under the base shall be equal to one another. Q. E. D.

[b] Ax. 3.

COROLLARY.

Hence every equilateral triangle is also equiangular.

PROPOSITION VI.

THEOREM.

If two angles of a triangle be equal to one another, the sides subtending the equal angles shall be equal to one another.

Let ABC be a triangle, having the angle ABC equal to the angle ACB. Then is the side AB equal to the side AC. For if AB be unequal to AC, one of them is greater. Let AB be the greater; and from the greater AB take away DB equal[a] to AC the less; and join DC. Therefore because DB is equal to AC; and BC common, the two DB, BC, will be equal to the two AC, CB, each to each; and the angle DBC equal to the angle ACB (by hypoth.). Whence the base DC is equal[b] to the base AB, and the triangle DBC equal to the triangle ACB, the less to the greater, which is absurd. Therefore the sides AB, AC, are not unequal. Whence they are equal. Wherefore if two angles of a triangle be equal to one another, &c. Q. E. D.

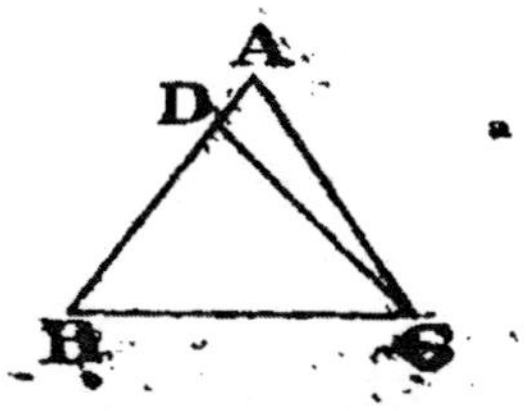

[a] 3. 1.

[b] 4. 1.

COROLLARY.

Hence every equiangular triangle is also equilateral.

PROPOSITION VII.

THEOREM.

On the same right line cannot be constituted two right lines equal to two other right lines, each to each, drawn to different points, to the same parts, and having the same extremes with the two right lines first drawn.

For if it be possible, let the two right lines AD, DB, be constituted upon the right line AB equal to two right lines AC, CB, each to each, *drawn* to different points, C and D, situated on the same side of the line AB, the lines AD, DB, having the same ends A, B, with the two first lines AC, CB; so that CA be equal to DA, both having the same end A; and CB be equal to DB, both having the same end B: for join the right line CD. Therefore because AC is equal to AD, the angle ACD will be equal to the angle ADC.[a] Whence the angle ADC is greater than the angle BCD. Wherefore the angle BDC will be much greater than the angle BCD. Again, because CB is equal to DB, the angle BDC will be equal to the angle BCD. But it has been shown to be much greater, which is impossible. Therefore on the same right line cannot be constituted two right lines, &c. Q. E. D.

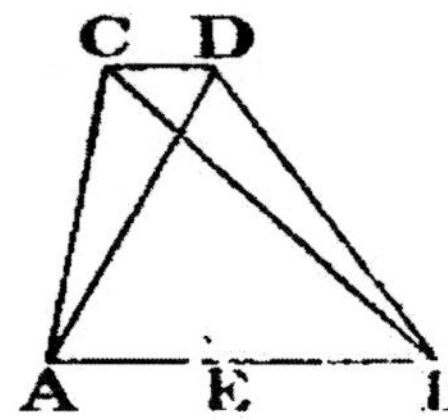

[a] 5. 1.

SCHOLIUM.

If D, one of the points C, D, be within the triangle ACB, a demonstration may be obtained by means of the latter part of the fifth proposition. For AC, AD, being drawn, the external angles ECD, FDC, which are under the base of the isosceles triangle ACD, will be equal to one another;[b] therefore the angle BDC will be greater than the angle ECD; whence the angle BDC will be much greater than the angle BCD; but because BD, BC, are equal, the angle BDC will be equal[b] to the angle BCD, a greater to a less, which is impossible.

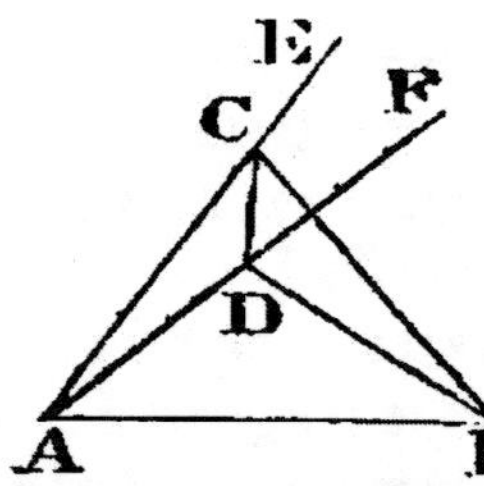

[b] 5. 1.

But if the point D be taken in either of them, AC, BC, the proposition is manifest; for so the whole AC would be equal to its part AD, or the whole BC equal to its part BD, which is impossible.

Dr. Simson, in his note to this proposition, says, he has thought proper to change its enunciation, "because (he adds) the literal translation from the Greek is extremely harsh, and difficult to be understood by beginners." Whatever difficulty learners may experience in this proposition, considered abstractedly, is easily removed by its exposition in the figure; and therefore it appears to me, that Dr. Simson has acted very injudiciously in altering its enunciation: and I perfectly agree with Taylor in saying, that it seems strange such liberties should be taken by one, who professes, in his preface, *to remove blemishes and restore the principal books of the Elements to their original accuracy.*

PROPOSITION VIII.

THEOREM.

If two triangles have two sides equal to two sides, each to each, and have their bases equal; the angle also, which is contained by the equal sides of the one triangle, shall be equal to the angle contained by the equal sides of the other.

Let there be two triangles ABC, DEF, which have two sides AB, AC, equal to two sides DE, DF, each to each, viz. AB equal to DE, and AC to DF; and they have the base BC equal to the base EF.

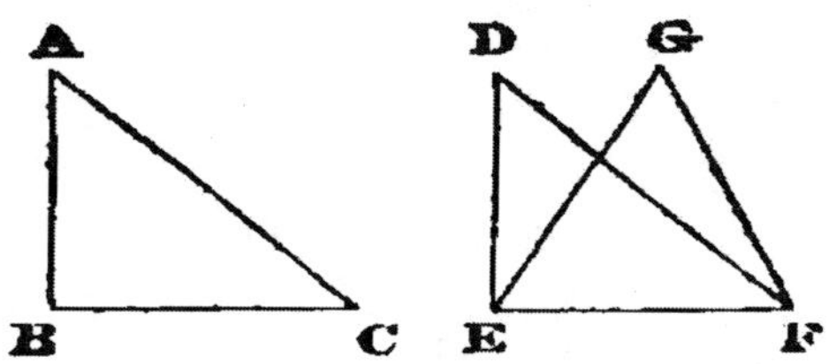

Then is the angle BAC equal to the angle EDF. For the triangle ABC being applied to the triangle DEF, and the point B being put on E; also the right line BC being applied to EF, the point C will coincide with the point F, because BC is equal to EF. Therefore BC coinciding with EF; BA, AC, will also coincide with ED, DF; for if the base BC coincide with the base EF; and the sides BA, AC, do not coincide with the sides ED, DF, but have a different situation, as EG, GF; then, on the same right line would be constituted two right lines equal to two other right lines, each to each, drawn to different points, to the same parts, and having the same extremes with the two right lines first drawn. But they cannot be so constituted as has been demonstrated.[a] Therefore if the base BC coincide with the base EF, the sides BA, AC, cannot but coincide with the sides ED, DF. Wherefore the angle BAC will also coincide with the angle EDF, and be equal to it. Therefore if two triangles have two sides equal to two sides, &c. Q. E. D.

[a] 7. 1.

Deduction from Euclid.

In an isosceles triangle, the right line drawn from the vertical angle bisecting the base is at right angles to the base.

Proclus has given a direct demonstration of this theorem, and is translated by Stone into his edition of the Elements, page 15. It is the converse of the fourth, although Euclid has not added so much as in that theorem, viz., *that the triangles and remaining angles are equal;* the reason is manifest, for the equality of the vertical angles being demonstrated, it follows, that all are equal to all by the fourth. Whence it was only necessary to demonstrate this, and assume the rest as consequents.

PROPOSITION IX.

Problem.

To bisect a given rectilineal angle, that is, to divide it into two equal parts.

Let BAC be the given rectilineal angle; it is required to bisect it. Take any point D in the right line AB, and from the
[a] 3. 1. line AC take AE equal[a] to AD, and
[b] 1. 1. DE being joined; upon it[b] describe the equilateral triangle DEF, and join AF. The angle BAC is bisected by the right line AF. For because AD is equal to AE, and AF common: the two DA, AF, are equal to the two EA, AF, each to each; and the base DF is equal
[c] 8. 1. to the base EF: therefore the angle[c] DAF is equal to the angle EAF. Wherefore the given rectilineal angle BAC is bisected by the right line AF. Q. E. F.

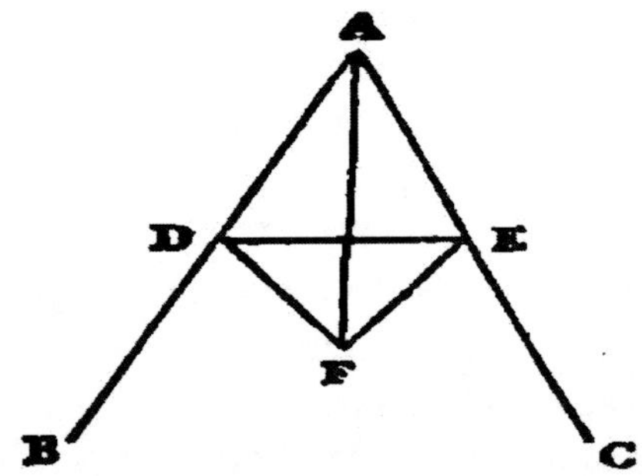

Deduction.

Divide a given rectilineal angle into any even number of equal parts.

PROPOSITION X.

Problem.*

To bisect a given finite right line, that is, to divide it into two equal parts.

Let AB be the given finite right line; it is required to bisect it. Upon
[a] 1. 1. it[a] describe the equilateral triangle
[b] 9. 1. ABC; and bisect the angle[b] ACB by the right line CD. The right line AB is bisected in the point D. For because AC is equal to CB, and CD common; the two AC, CD, are equal to the two BC,

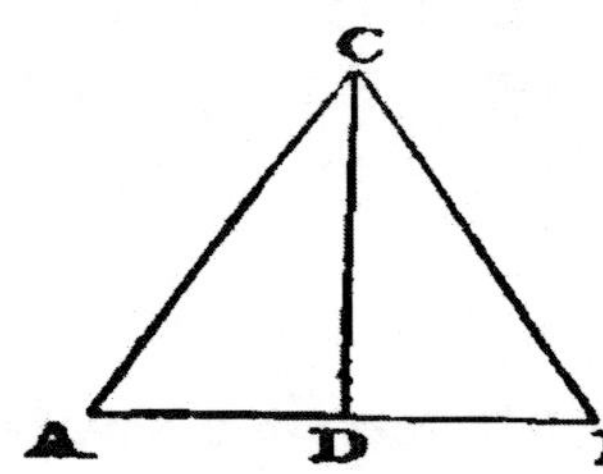

* A given finite right line may also be bisected by means of the construction to the first proposition of this book, and joining the common sections of the circles.

CD, each to each; and the angle ACD is equal to the angle BCD: therefore the base AD is equal[c] to the base BD. And consequently the finite right line AB is bisected in the point D. Q. E. F. [c] 4. 1.

Deduction.

From the vertex of a given scalene triangle, to draw, to the base, a straight line which shall exceed the less of the two sides, as much as it is itself exceeded by the greater.

PROPOSITION XI.

PROBLEM.

To a given right line, from a given point in it; to draw a right line at right angles to the former.

Let AB be the given right line, and C a given point in it, it is required to draw from the point C a right line at right angles to AB. Take any point D in AC, and make CE equal to CD,[a] and upon DE describe the equilateral triangle FDE,[b] and join FC. The right line FC is drawn at right angles to the given right line AB from the point C given in it. For because DC is equal to CE, and FC common; the two DC, CF, will be equal to the two EC, CF, each to each; and the base DF is equal to the base EF; wherefore the angle DCF is equal to the angle ECF, and they are adjacent angles. But when a right line standing upon a right line makes the adjacent angles equal to one another, each of them is a right angle: therefore each of the angles DCF, ECF, is a right angle. Wherefore the right line FC is drawn at right angles to the given right line AB, from the point C given in it. Q. E. F.

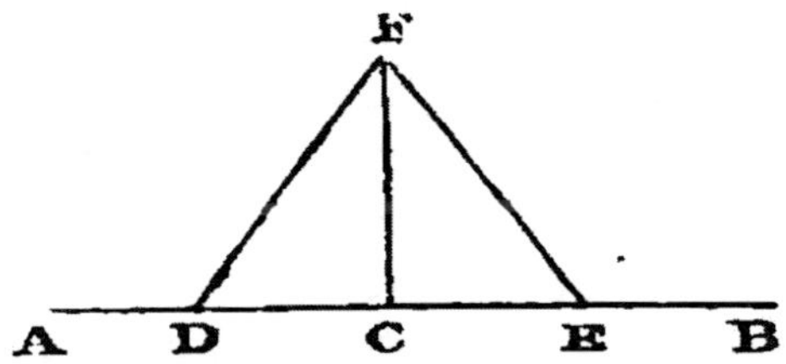

[a] 3. 1.
[b] 1. 1.

Deductions.

1. Describe a circle which shall pass through three given points which are not in the same right line.
2. In a right line given in position, but indefinite in

length, to find a point, which shall be equidistant from each of two given points, either on contrary sides, or both on the same side of the given line, and in the same plane with it; but not situated in a perpendicular to it.

PROPOSITION XII.*

Problem.

Upon a given infinite right line from a given point which is without it; to draw a perpendicular right line.

Let AB be the given infinite right line, and C a given point which is without it, it is required to draw upon the given infinite right line AB a perpendicular from the given point C, which is without it. Take any point D upon the other side of AB, and from the centre C at the distance CD describe the circle EDG; and

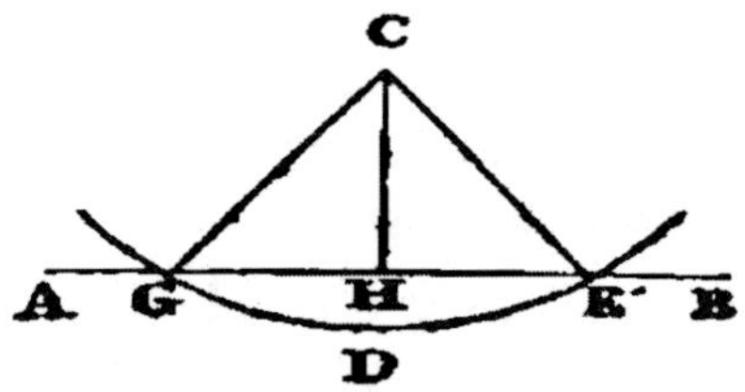

[a] 10. 1. bisect GE in H;[a] and join CG, CH, CE. The perpendicular CH is drawn upon the given infinite right line AB from the point C, which is without it.

For because GH is equal to HE, and HC common, the two GH, HC, are equal to the two EH, HC, each to each; and the base CG is equal to the base CE. Therefore the
[b] 8. 1. angle CHG is equal to the angle CHE,[b] and they are adjacent angles. But when a right line standing upon another right line makes the adjacent angles equal to one another, each of them is a right angle, and the right line standing upon the other is called a perpendi-
[c] 10. Def. 1. cular;[c] wherefore upon a given infinite right line AB, from the given point C, which is without it, CH has been drawn perpendicular. Q. E. F.

Deduction.

Given the vertex of a triangle, the perpendicular from the vertex to the base and also the base, to construct the triangle.

---

* Euclid did well in proposing an infinite right line, for otherwise the given point might be situated in a direct position with the given line, and consequently the problem would not succeed.

PROPOSITION XIII.

THEOREM.

When a right line standing upon a right line makes angles, these are either two right angles, or are equal to two right angles.

For let a certain right line AB standing upon the right line CD make the angles CBA, ABD. The angles CBA, ABD, are either two right angles, or are equal to two right angles. For if CBA be equal to ABD, they are right angles;[a] but if less, draw from the point B, BE at right angles to DC;[b] the angles CBE, DBE, are therefore two right angles; and because CBE is equal to the two CBA, ABE, add EBD, which is common; therefore the two angles CBE, EBD, are equal to the three angles CBA, ABE, EBD.[c] Again, because the angle DBA is equal to the two DBE, EBA, and ABC, which is common; therefore the two angles DBA, ABC, are equal to the three DBE, EBA, ABC. But it was shown that the angles CBE, EBD, are equal to the same three, and things which are equal to the same are equal to one another; therefore the angles CBE, EBD, are equal to DBA, ABC; but CBE, EBD, are two right angles; therefore the angles DBA, ABC, are equal to two right angles. Therefore when a right line standing upon a right line, &c. Q. E. D.

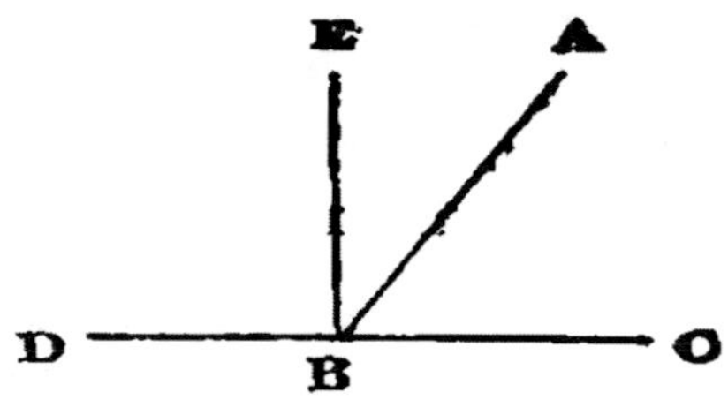

[a] Def. 10.

[b] 11. 1.

[c] Ax. 2.

PROPOSITION XIV.

THEOREM.

If to a certain right line, and to a point in it, two right lines not placed towards the same parts, make the adjacent angles equal to two right angles; the right lines will be in one and the same straight line.

For to a certain right line AB, and to a point in it B, let there be two right lines BC, BD, not placed toward the same parts, make the adjacent angles ABC, ABD, equal to two right angles. Then the line BD is in the same straight line with CB; for if BD is not in the same straight line

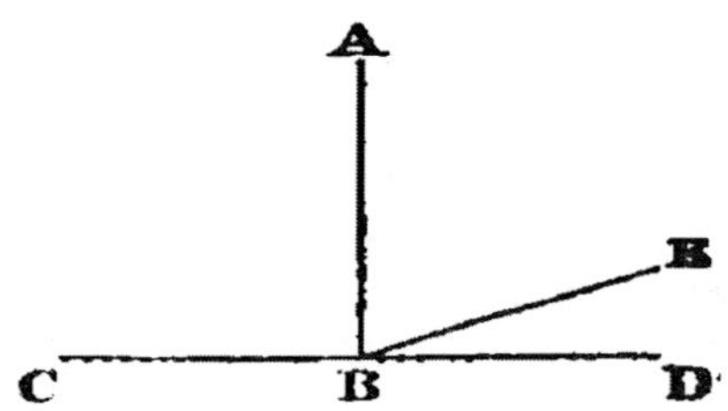

with CB, let BE be in the same straight line with it. Therefore, because the right line AB stands upon the right line CBE, the angles ABC, ABE, are equal to two
[a] 13. 1. right angles.[a] But also the angles ABC, ABD, are equal to two right angles. Therefore the angles CBA, ABE, will be equal to the angles CBA, ABD. Take away ABC, common to both. Therefore the remaining angle ABE is
[b] Ax. 3. equal to the remaining angle ABD,[b] the less to the greater, which is impossible. Therefore BE will not be in the same straight line with BC. In like manner we may show, that not any other can be except BD. Therefore BD will be in a right line with BC. If therefore to a certain right line, &c. Q. E. D.

PROPOSITION XV.

Theorem.*

If two right lines cut one another, they will make the vertical angles equal to one another.

For let the two right lines AB, CD, cut one another in the point E. Then the angle AEC is equal to the angle DEB; and the angle CEB equal to the angle AED. For because the right line AE standing upon the right line CD makes the angles CEA, AED; these will be
[a] 13. 1. equal[a] to two right angles. Again, because the right line DE standing upon the right line AB makes the angles AED, DEB; the angles AED, DEB, will be equal to two right angles. But it was shown that the angles CEA, AED, are equal to two right angles. Therefore the angles CEA, AED, are equal to the angles AED, DEB. Take away the common angle AED. Therefore the remaining angle
[b] Ax. 3. CEA is equal[b] to the remaining angle BED. In like manner it may be shown that the angles CEB, AED, are equal. If therefore two right lines cut one another, &c. Q. E. D.

* This proposition was discovered by Thales, according to the account given by Eudemus; and the corollaries ought, I think, to be rather placed after the thirteenth, as they are the natural consequences of that theorem, upon which the demonstration of the fifteenth entirely depends.

COROLLARIES.

1. From this it is manifest, if two right lines cut one another, they make the angles at the point of section equal to four right angles.

2. All the angles placed around one point are equal to four right angles.

PROPOSITION XVI.

THEOREM.

One side of any triangle being produced, the exterior angle is greater than either of the interior and opposite angles.

Let ABC be a triangle, and let one of its sides BC be produced to D. The exterior angle ACD is greater than either of the interior and opposite angles; namely, the angles CBA and BAC. Bisect AC in E, and BE being joined, produce it to F, and make EF equal to BE; join also FC, and produce AC to G. Therefore because AE is equal to EC, and BE to EF, the two AE, EB, are equal to the two CE, EF, each to each; and the angle AEB is equal[a] to the angle FEC, for they are vertically opposite. Therefore the base AB is equal[b] to the base FC; and the triangle AEB to the triangle FEC; also the remaining angles to the remaining angles each to each, to which the equal sides are opposite. Therefore the angle BAE is equal to the angle ECF. But the angle ACD is greater than ECF. Therefore the angle ACD is greater than the angle BAE. In like manner it may be shown, the right line BC being bisected, that the angle BCG, that is, ACD, is greater than the angle ABC. Therefore one side of a triangle being produced, &c. Q. E. D.

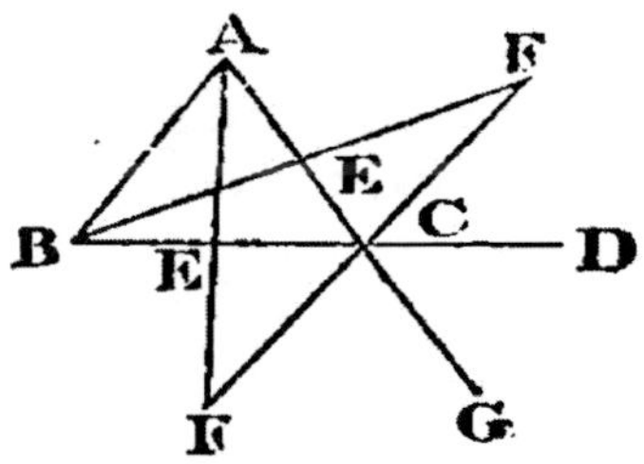

[a] 15. 1.

[b] 4. 1.

PROPOSITION XVII.

THEOREM.

Two angles of every triangle, howsoever taken, are less than two right angles.

Let ABC be a triangle. Two angles of the triangle ABC, howsoever taken, are less than two right angles. Produce BC to D; and because the exterior angle ACD

[a] 16. 1. of the triangle ACB is greater[a] than the interior and opposite angle ABC; add ACB, which is common. Therefore the angles ACD, ACB, are greater than the angles ABC, ACB. But ACD,
[b] 13. 1. ACB, are equal[b] to two right angles, whence ABC, BCA, are less than two right angles. In like manner, we may demonstrate also that the angles BAC, ACB, and also CAB, ABC, are less than two right angles. Therefore two angles of every triangle, &c. Q. E. D.

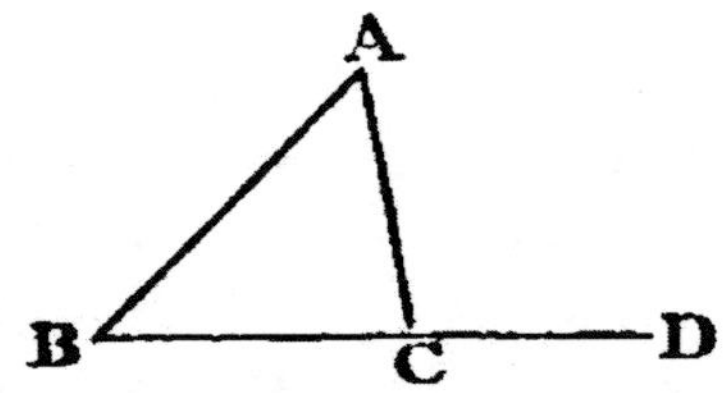

PROPOSITION XVIII.

Theorem.

The greater side of every triangle subtends the greater angle.

Let ABC be a triangle having the side AC greater than the side AB. The angle ABC is greater than the angle BCA. For because AC is greater than AB, make AD equal to AB, and join BD. And because ADB is the exterior angle, it will be greater than the
[a] 16. 1. interior and opposite angle DCB.[a] But
[b] 5. 1. ADB is equal to ABD,[b] because the side AB is equal to the side AD; therefore the angle ABD is greater than the angle ACB. Wherefore ABC will be much greater than ACB. Therefore the greater side of every triangle, &c. Q. E. D.

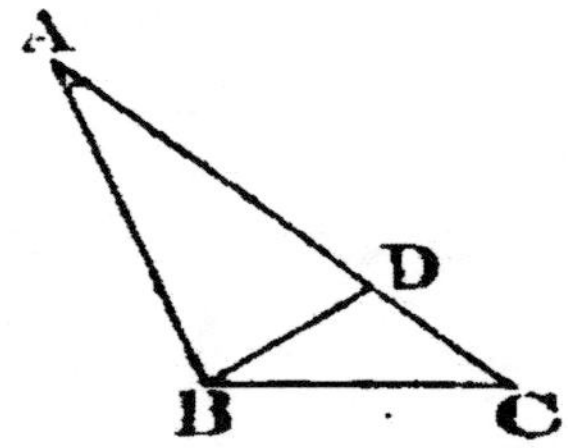

PROPOSITION XIX.

Theorem.

The greater angle of every triangle subtends the greater side.

Let ABC be a triangle having the angle ABC greater than the angle BCA. The side AC is greater than the side AB. For if AC is not greater, it is either equal to it or less; but it is not equal; for then the angle ABC would be
[a] 5. 1. equal to the angle ACB;[a] but it is not. Therefore AC is not equal to AB. But neither is it less; for then the angle ABC
[b] 18. 1. would be less than the angle ACB,[b] but it is not. There-

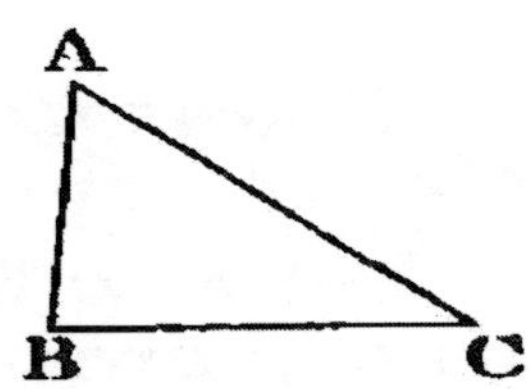

fore AC is not less than AB. And it was shown that it is not equal. Whence AC is greater than AB. Therefore the greater angle, &c. Q. E. D.

Deduction.

If from a given point there be drawn any number of right lines to another right line given in position, the perpendicular is the shortest right line, and that which is nearer to the perpendicular is less than that which is more remote, and there can only be drawn two right lines which are equal to one another, one on each side of the perpendicular.

PROPOSITION XX.

THEOREM.

Two sides of every triangle, howsoever taken, are greater than the third side.

For let ABC be a triangle, any two sides of the triangle ABC, howsoever taken, are greater than the third side; viz. the sides BA, AC, are greater than the side BC; the sides AB, BC, greater than AC; and the sides BC, CA, are greater than AB. For produce BA to the point D, and make CA equal to AD,[a] and join DC. Therefore because DA is equal to AC, the angle ADC will be equal to the angle ACD.[b] But the angle BCD is greater than the angle ACD; therefore the angle BCD is greater than the angle ADC. And because DCB is a triangle having the angle BCD greater than the angle BDC, and the greater side subtends the greater angle;[c] the side DB will be greater than the side BC; but DB is equal to BA, AC; wherefore the sides BA, AC, will also be greater than BC. In like manner we may show that the sides AB, BC, are greater than AC; and the sides BC, CA, greater than AB. Therefore two sides of every triangle, &c.

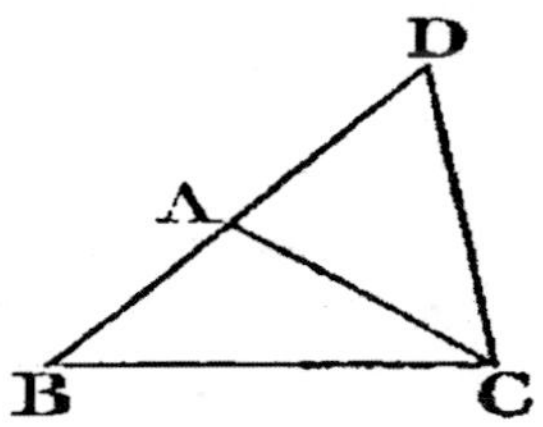

[a] 3. 1.
[b] 5. 1.
[c] 19. 1.

Deductions.

1. The difference of any two sides of a triangle is less than the third side.
2. Two sides of a triangle are together greater than twice the line from the vertex bisecting the base.

PROPOSITION XXI.

Theorem.

If from the ends of one side of a triangle two right lines be drawn within it, these will be less than the other two sides of the triangle, but will contain a greater angle.

For from the ends BC in one of the sides BC of the triangle ABC, let two right lines BD, DC, be drawn within it. The sides BD, DC, are less than the two sides BA, AC, of the triangle, but contain the angle BDC greater than the angle BAC. Produce BD to E; and because two sides of every triangle are

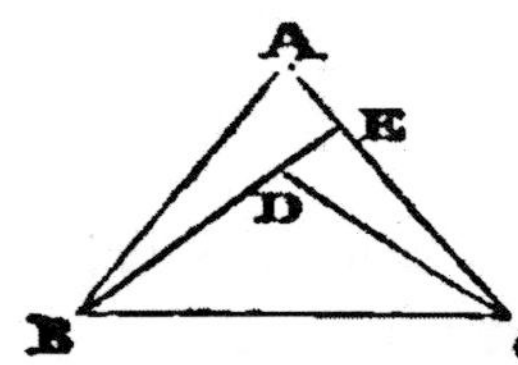

[a] 20. 1. greater than the third side,[a] the two sides BA, AE, of the triangle ABE, are greater than BE. Add EC, which is common. Therefore BA, AC, are greater

[b] Ax. 4. than BE, EC.[b] Again, because CE, ED, two sides of the triangle CED, are greater than CD, add DB, which is common: wherefore CE, EB, are greater than CD, DB. But it has been shown that BA, AC, are greater than BE, EC; much more then are BA, AC, greater than BD, DC. Again, because the exterior angle of every triangle is greater than the interior and opposite

[c] 16. 1. angle,[c] the exterior angle BDC of the triangle CDE will be greater than the interior and opposite angle CED. For the same reason, the exterior angle CEB of the triangle ABC is greater than BAC. But the angle BDC was shown to be greater than the angle CEB; much more then will the angle BDC be greater than the angle BAC. Wherefore if from the ends, &c. Q. E. D.

PROPOSITION XXII.

Problem.

To make a triangle of three right lines which shall be equal to three given right lines. But any two of these lines, howsoever taken, will be greater than the third; because any two sides of a triangle, howsoever taken, are greater than the third side.

Let A, B, C, be three given right lines, two of which, howsoever taken, are greater than the third, viz. A, B, greater than C; A, C, greater than B; and B, C, greater than A. It is required to make a triangle whose three right lines are equal to three given right lines A, B, C. Take any right line DE terminated at D, but unlimited towards E, and make DF equal to A[a]; FG equal to B, and GH equal to C, and from the centre F at the distance FD, describe the circle DKL;[b] and again from the centre G at the distance GH describe another circle KLH, and join KF, KG. The triangle KFG is made, whose three right lines are equal to A, B, C. For because F is the centre of the circle DKL, FD will be equal to FK;[c] but FD is equal to A; therefore FK is also equal to A. Again, because the point G is the centre of the circle LKH, GH will be equal to GK, but GH is equal to C; therefore GK will be equal to C. And FG is equal to B. Therefore the three right lines KF, FG, GK, are equal to the three A, B, C. Therefore the triangle KFG has been made whose three sides KF, FG, GH, are equal to the three right lines A, B, C. Q. E. F.

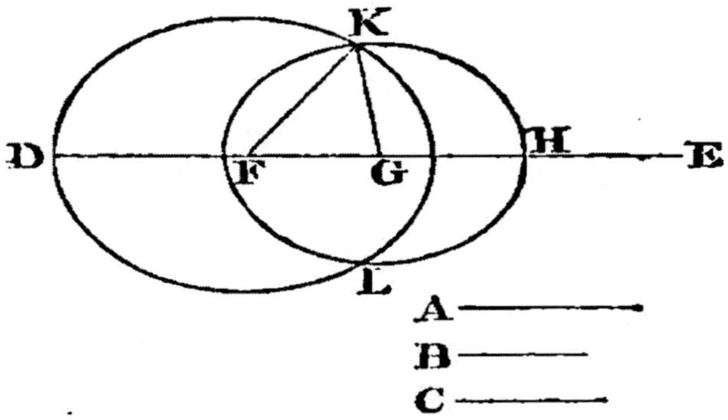

[a] 3. 1.

[b] 3 Post.

[c] Def. 15.

Deduction.

It is required to construct a rectilineal figure of any number of sides, having given the length of each side of all the triangles into which the figure is divided.

PROPOSITION XXIII.

Problem.*

To a given right line, and to a given point in it, to make a rectilineal angle equal to a given rectilineal angle.

Let AB be the given right line, and A the given point in it; also DCE the given rectilineal angle. It is required therefore to the given right line AB, and to the point A in it, to make a rectilineal angle equal to the given rectilineal angle DCE. Take in each of them CD, CE, the

---

* Apollonius has given a more simple and easy solution of this Problem; but as the demonstration of that method requires the assistance of Prop. 27, Book 3, Euclid could not introduce it into this place; since a well connected chain of consequences, and an uniform assumption of principles, previously demonstrated, were the main objects of Euclid's plan, and which, I think, constitute the beauty and superiority of his Elements.

points DE; join DE, and make the triangle AFG whose three right lines are equal to the three right lines CD,
[a] 22. 1. DE, EC.[a] So that AF be equal to CD, AG to CE, and FG to DE. Therefore because the two DC, CE, are equal to the two FA, AG, each to each, and the base DE equal to the base FG, the angle
[b] 8. 1. DCE will be equal to the angle FAG.[b] Therefore to a given right line AB, and to a point in it, a rectilineal angle has been placed equal to the given rectilineal angle DCE. Q. E. F.

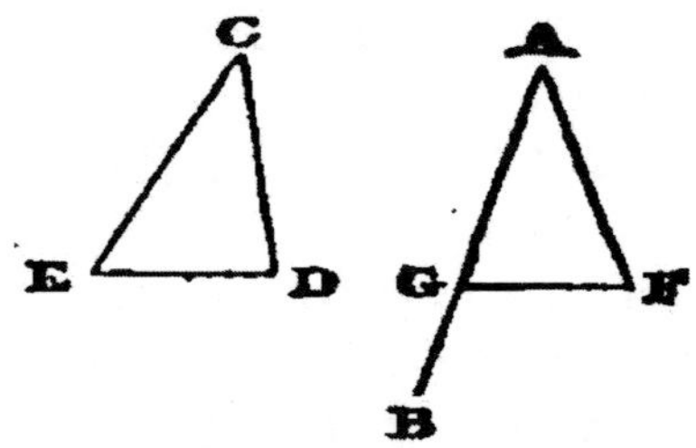

PROPOSITION XXIV.

Theorem.*

If two triangles have two sides equal to two sides, each to each, but the angle contained by the equal sides of the one greater than the angle contained by the equal sides of the other; then shall the base of that which has the greater angle be greater than the base of the other.

Let ABC, DEF, be two triangles, which have the two sides AB, AC, equal to the two sides DE, DF, each to each, viz. the side AB equal to the side DE, and the side AC equal to DF. But the angle BAC greater than the angle EDF. The base BC will be greater than the base EF. For because the angle BAC is greater than the angle EDF; to the right line DE and to the point D in it, make the angle
[a] 23. 1. EDG equal to BAC;[a] also put DC equal to either AC, or
[b] 3. 1. DF,[b] and join GE, FG. Therefore because AB is equal to DE, and AC to DG; the two BA, AC, are equal to the two ED, DG, each to each; and the angle BAC is equal to the angle EDG; therefore the base BC is equal to the
[c] 4. 1. base EG.[c] Again, because DG is equal to DF, the angle
[d] 5. 1. DFG is equal to the angle DGF;[d] therefore the angle DFG will be greater than the angle EGF, much more will the angle EFG be greater than the angle EGF. And

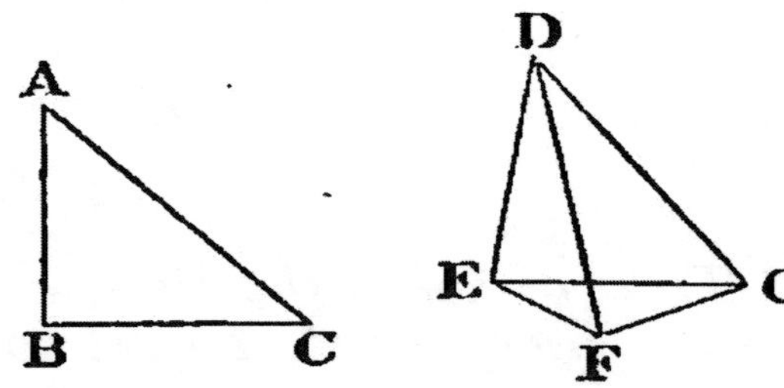

* The student is not to conclude from hence that the area of that triangle which has the greater base is greater than the area of the other, for it can be clearly proved, by means of some following propositions, that its area may be either equal to the other triangle, or less than it.

because EFG is a triangle having the angle EFG greater than the angle EGF, also the greater side subtends the greater angle,[e] the side EG will be greater than the side EF. But the side EG is equal to the side BC. Therefore BC will also be greater than EF. If therefore two triangles have two sides, &c. Q. E. D. [e] 19. 1.

PROPOSITION XXV.

THEOREM.*

If two triangles have two sides equal to two sides each to each; but the base of the one greater than the base of the other; then shall the angle contained by the equal sides of the one be greater than the angle which is contained by the equal sides of the other.

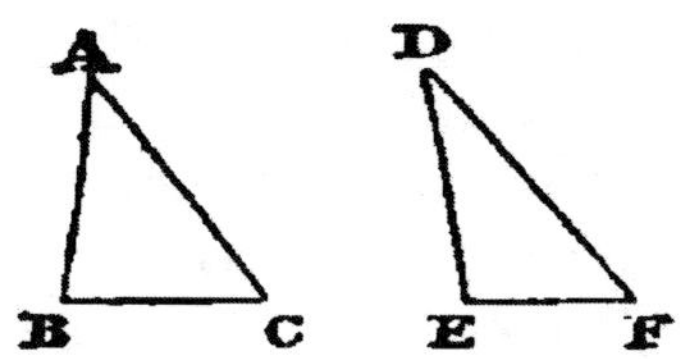

Let ABC, DEF, be two triangles, which have the two sides AB, AC, equal to the two sides DE, DF, each to each, viz. the side AB equal to the side DE, and the side AC to the side DF. But the base BC greater than the base EF. The angle BAC is greater than the angle EDF. For if it be not greater, it is either equal or less. But the angle BAC is not equal to the angle EDF; for then the base BC would be equal to the base EF.[a] But it is not. [a] 4. 1. The angle BAC is not therefore equal to the angle EDF. But neither is it less; for then the base BC would also be less than the base EF.[b] But it is not. [b] 24. 1. Therefore the angle BAC is not less than the angle EDF. And it was shown that it is not equal. Therefore the angle BAC is greater than EDF. If therefore two triangles, &c. Q. E. D.

PROPOSITION XXVI.

THEOREM.

If two triangles have two angles equal to two angles, each to each, and one side equal to one side, either the side which is adjacent to the equal angles, or the side which subtends one of the equal angles, they will have the remaining sides equal to the remaining sides each to each, and the remaining angle to the remaining angle.

Let ABC, DEF, be two triangles, which have the two angles ABC, BCA, equal to the two angles DEF, EFD,

* Direct demonstrations of this are given by Menelaus, Alexandrinus, and Hero.

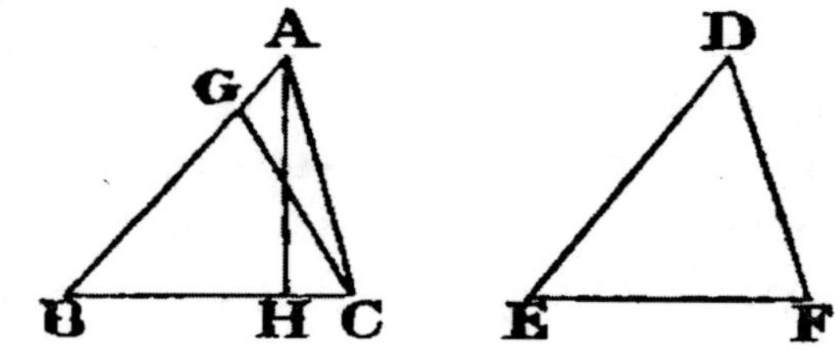

each to each, viz. the angle ABC equal to the angle DEF, and the angle BCA equal to the angle EFD. Let them also have one side equal to one side, and first that which is adjacent, to the equal angles, viz. the side BC to the side EF. They will also have the remaining sides equal to the remaining sides, each to each, viz. the side AB to the side DE, and the side AC to DF, and the remaining angle BAC equal to the remaining angle EDF. For if AB is unequal to DE, one of them is the greater. Let AB be the greater, and make GB equal to DE; and join GC. Therefore because BG is equal to DE, and BC to EF; the two GB, BC, are equal to the two DE, EF, each to each; and the angle GBC is equal to the angle DEF. Therefore the base GC is equal to the base DF,[a] and the triangle GBC to the triangle DEF; also the remaining angles equal to the remaining angles, each to each, to which the equal sides are opposite. Therefore the angle GCB is equal to the angle DFE; but the angle DFE is equal to the angle BCA; wherefore also the angle BCG is equal to the angle BCA, the less to the greater; which is impossible. Therefore AB is not unequal to DE; that is, it is equal to it. But BC is equal to EF. Therefore the two AB, BC, are equal to the two DE, EF, each to each, and the angle ABC is equal to the angle DEF. Therefore the base AC is equal to the base DF, and the remaining angle BAC is equal to the remaining angle EDF; but let the sides, which subtend the equal angles, be equal to one another, as AB to DE. Then again the remaining sides are equal to the remaining sides; viz. AC is equal to DF, also BC to EF; and, as before, the remaining angle BAC is equal to the remaining angle EDF. For if BC be unequal to EF, one of them is the greater. Let BC be the greater, if it be possible, and make BH equal to EF, and join AH. Wherefore because BH is equal to EF, and AB to DE; the two AB, BH, are equal to the two DE, EF, each to each, and they contain equal angles; therefore the base AH is equal to the base DF; and the triangle ABH to the triangle DFE; also the remaining angles will be equal to the remaining angles, each to each, to which the equal sides are opposite. Therefore the angle BHA is equal to the angle EFD. But EFD is

[a] 4. 1.

equal to the angle BCA.[b] And therefore the angle BHA is equal to the angle BCA; the exterior angle BHA of the triangle AHC is equal to the interior and opposite angle BCA, which is impossible.[c] Wherefore BC is not unequal to EF; that is, it is equal to it. But AB is equal to DE. Therefore the two AB, BC, are equal to the two DE, EF, each to each, and they contain equal angles. Wherefore the base AC is equal to the base DF, and the triangle BAC to the triangle DEF; also the remaining angle BAC is equal to the remaining angle EDF. If therefore two triangles, &c. Q. E. D.

[b] By hyp.
[c] 16. 1.

Deductions.

1. To draw a right line through a given point so as to make equal angles with two right lines given in position.

2. If any two triangles have the three angles of the one respectively equal to the three angles of the other; also if the perpendiculars from the vertical angles to the bases be equal, then shall the three sides of the one triangle be equal to the three sides of the other, viz. those which are opposite to the equal angles.

PROPOSITION XXVII.

THEOREM.*

If a right line falling upon two right lines makes the alternate angles equal to one another, the right lines will be parallel.

Let the right line EF falling upon the two right lines AB, CD, make the alternate angles AEF, EFD, equal to one another; the right line AB is parallel to CD. For if it is not parallel, AB, CD, being produced will meet either towards the parts B, D, or towards the parts A, C. Let them be produced, and meet towards the parts B, D, in the point G. Therefore the exterior angle AEF of the triangle GEF is greater than the interior and opposite angle EFG;[a] but it is also equal;[b] which is impossible. Therefore AB, CD, being produced, do not meet towards the parts BD. In like manner we may demonstrate they do not meet

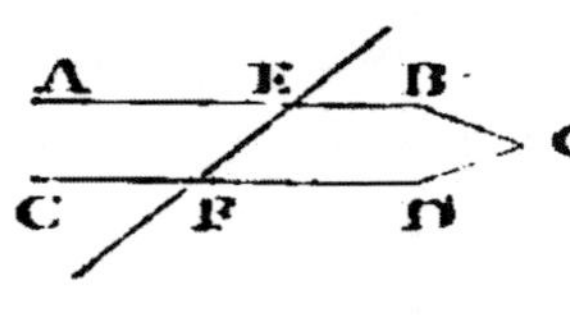

[a] 16. 1.
[b] By hyp.

* The student must understand that the lines are in the same plane, otherwise the alternate angles might be equal; and the lines might or might not be parallel, as the commentary of Proclus upon this proposition fully evinces.

towards the parts A, C. But those right lines which being produced meet towards neither parts are parallel. Therefore AB is parallel to CD. Wherefore a right line, &c. Q. E. D.

Deduction.

If ABCD be a parallelogram, and BH be equal to DG, the triangle BFH is equal to the triangle DGE; also the sides and angles of the one respectively equal to the sides and angles of the other.

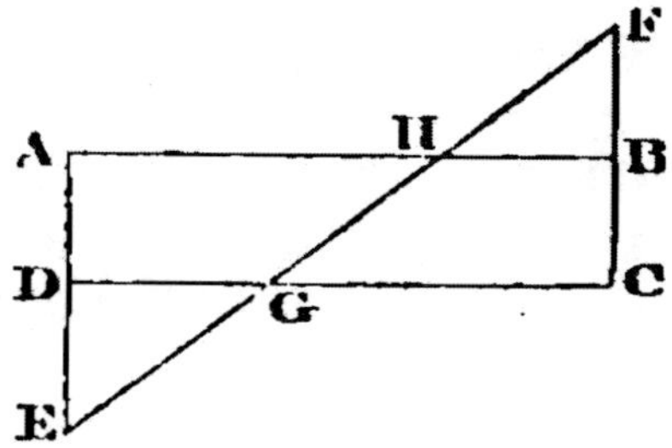

PROPOSITION XXVIII.

THEOREM.

If a right line falling upon two right lines make the exterior angle equal to the interior and opposite angle towards the same parts, or the interior angles towards the same parts equal to two right angles; the right lines will be parallel to one another.

Let the right line EF falling upon the two right lines AB, CD, make the exterior angle EGB equal to the interior and opposite angle GHD, or the interior angles towards the same parts BGH, GHD, equal to two right angles. The right line AB is parallel to the right line CD. For because the angle EGB is equal to the angle GHD,[a] and the angle EGB to the angle AGH,[b] the angle AGH will also be equal to the angle GHD; and they are alternate angles. Therefore AB is parallel to CD.[c] Again, because the angles BGH, GHD, are equal to two right angles,[d] and the angles AGH, BGH, are equal to two right angles;[e] therefore the angles AGH, BGH, will be equal to the angles BGH, GHD. Take away BGH, which is common. Therefore the remainder AGH is equal to the remainder GHD; and they are alternate angles. Therefore AB will be parallel to CD. If therefore a right line, &c. Q. E. D.

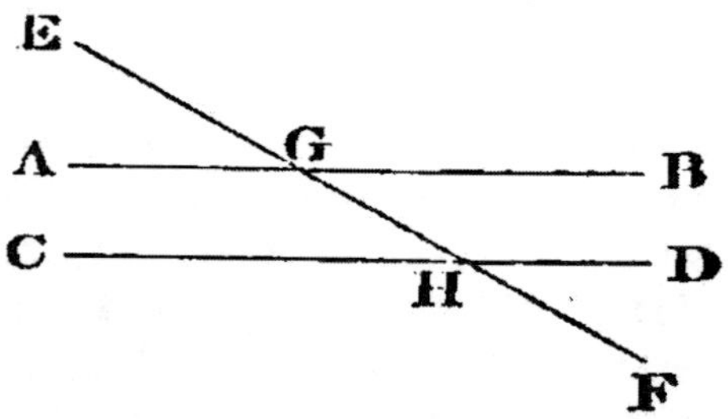

[a] By hyp.
[b] 15. 1.
[c] 27. 1.
[d] By hyp.
[e] 13. 1.

PROPOSITION XXIX.*

THEOREM.

If a right line falls upon two parallel right lines, it will make the alternate angles equal to one another, the exterior angle equal to the interior and opposite angle towards the same parts, and the interior angles towards the same parts equal to two right angles.

Let the right line EF fall upon the two right lines AB, CD. The alternate angles AGH, GHD, are equal to one another, the exterior angle EGB towards the same parts, equal to the interior and opposite angle GHD. And the interior angles BGH, GHD, towards the same parts equal to two right angles. For if AGH be unequal to GHD, one of them is the greater. Let AGH be the greater. And

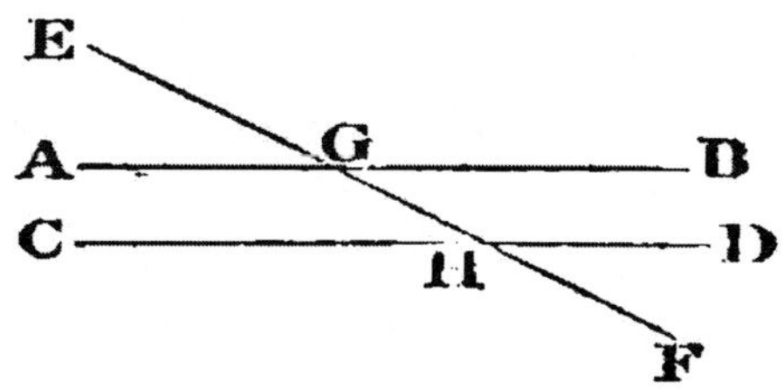

because the angle AGH is greater than the angle GHD, add BGH, which is common. Therefore the angles AGH, BGH, are greater than the angles BGH, GHD. But the angles AGH, BGH, are equal to two right angles.[a] Therefore the angles BGH, GHD, are less than two right angles. But right lines which with another right line falling upon them make the adjacent angles less than two right angles, do meet, if produced far enough.[b] Therefore the right lines AB, CD, produced far enough, will meet. But they do not meet, since they are parallel. Therefore the angle AGH is not unequal to the angle GHD; wherefore it is equal. But the angle AGH is equal to the angle EGB.[c] Therefore EGB will be equal to GHD; add BGH, common to both. Therefore the angles EGB, BGH, are equal to the angles BGH, GHD; but EGB, BGH, are equal to two right angles. Therefore BGH, GHD, will be equal to two right angles. If, therefore, a right line, &c. Q. E. D.

[a] 13. 1.
[b] Ax. 12.
[c] 15. 1.

---

* This and the preceding 27th Proposition show the excellency of Euclid's definitions of parallels, and its superiority to many others given by the moderns; for he here employs the negative property of these lines with great success, and the addition of their being always at the same perpendicular distance from each other would have been useless, as it is not wanted in any part of the Elements.

Deduction.

Trisect a right angle; that is, to divide it into three equal parts, trisect any rectilineal angle which is an even aliquot part of a right angle.

PROPOSITION XXX.

Theorem.

Right lines which are parallel to the same right line, are parallel to each other.

Let AB, CD, be each of them parallel to EF; then AB, CD, are parallel to one another. Let GK, a right line, fall upon them. And because the right line GK falls upon the parallel right lines AB, EF, the angle AGH is equal to the angle

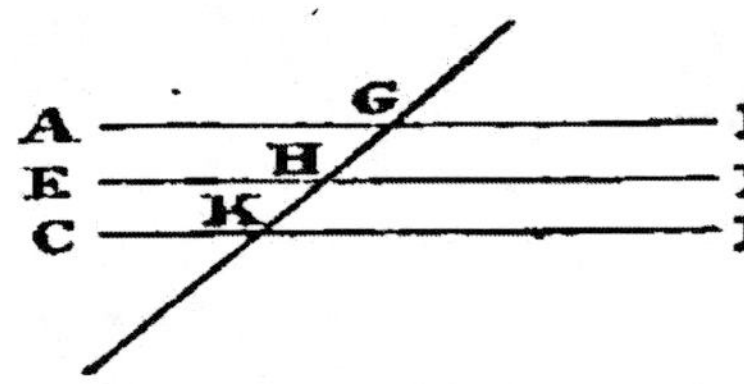

[a] 29. 1. GHF.[a] Again, because the right line GK falls upon the parallel right lines EF, CD, the angle GHF is equal to the angle GKD. But it was shown the angle AGK is also equal to the angle GHF; therefore AGK will also be equal to GKD; and they are alternate angles. Therefore AB is parallel

[b] 27. 1. to CD.[b] Wherefore the right lines, &c. Q. E. D.

PROPOSITION XXXI.

Problem.

Through a given point to draw a right line parallel to a given right line.

Let A be a given point, and BC a given right line. It is required through the point A to draw a right line parallel to the right line BC. Take in BC any point D, and join AD; place at the right line DA, and at the point A in it, the angle DAE

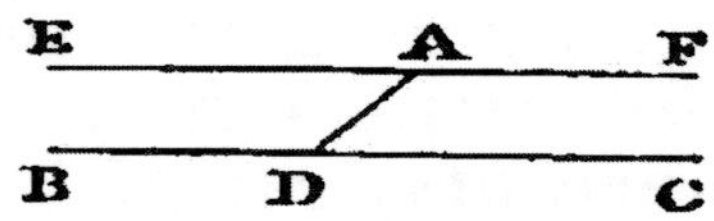

[a] 23. 1. equal to the angle ADC;[a] and produce the right line AF in a straight line with EA. For because the right line AD falling upon the right lines BC, EF, makes the alternate angles EAD, ADC, equal to

[b] 27. 1. one another; EF shall be parallel to BC.[b] Therefore, through the point A, a right line EAF has been drawn parallel to a given right line BC. Q. E. F.

Deductions.

1. To draw to a right line from a given point without it another right line, which shall make an angle equal to a given rectilineal angle.

2. From a given isosceles triangle to cut off a trapezium, which shall have the same base as the triangle, and shall have its three remaining sides equal to each other.

PROPOSITION XXXII.*

THEOREM.

One side of any triangle being produced, the exterior angle is equal to the two interior and opposite angles, and the three interior angles of a triangle are equal to two right angles.

Let ABC be a triangle; and let one of its sides BC be produced to D. The exterior angle ACD is equal to the two interior and opposite angles CAB, ABC; and the three interior angles of the triangle, viz. ABC, BCA, CAB, are equal to two right angles. For through the point C draw CE parallel to the right line AB;[a] and because AB is parallel to CE, and AC falls upon them, the alternate angles BAC, ACE, are equal to one another.[b] Again, because AB is parallel to CE, and the right line BD falls upon them, the exterior angle ECD is equal to the interior and opposite angle ABC. But the angle ACE was shown to be equal to the angle BAC. Wherefore the whole exterior angle ACD is equal to the two interior and opposite angles BAC, ABC. Add ACB, which is common: therefore the angles ACD, ACB,

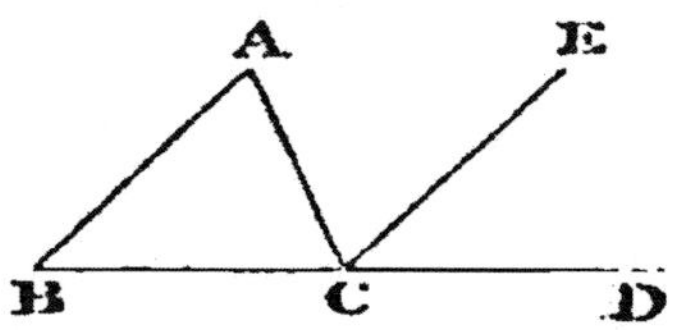

[a] 31. 1.

[b] 29. 1.

---

* The three angles of a triangle may be demonstrated to be equal to two right angles without the aid of the first part of the proposition, as Eudemus relates was done by the Pythagoreans, in the following manner. Let there be a triangle ABC, and let there be drawn through the point A, a line DE parallel to BC. Because therefore the right lines DE, BC, are parallel, the alternate angles are equal. Hence the angle DAB is equal to the angle ABC, and the angle EAC equal to the angle ACB. Let the common angle BAC be added. The angles, therefore, DAB, BAC, CAE; that is, the angles DAB, BAE, and that is two right angles, are equal to the three angles of the triangle.

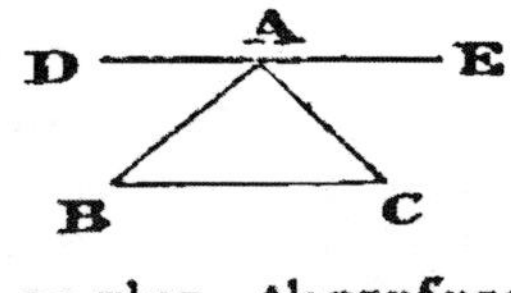

are equal to the three angles, ABC, BCA, CAB; but the
c 13. 1. angles ACD, ACB, are equal to two right angles.[c] Therefore the angles ACB, CBA, CAB, will also be equal to two right angles. Therefore if one side of every triangle, &c. Q. E. D.

Deductions.

1. The angle at the base of an isosceles triangle is equal to half the difference between the vertical angle and two right angles.

2. The difference of the angles at the base of any triangle is double the angle contained by the line bisecting the vertical angle and another drawn from the vertex perpendicular to the base.

3. Given the difference of the angles at the base of a triangle, the perpendicular drawn from the vertex to the base, and one of the segments made by the perpendicular, to construct the triangle.

4. To construct a triangle, which shall have its three sides taken together equal to a given finite right line, and its three angles equal to three given angles, each to each; the three given angles being together equal to two right angles.

PROPOSITION XXXIII.*

Theorem.

Right lines, which join equal and parallel right lines towards the same parts, are themselves equal and parallel.

Let AB, CD, be equal and parallel, and let the right lines AC, BD, join them towards the same parts. AC, BD, are equal and parallel to one another. Draw BC, and because AB is parallel to CD, and BC falls upon them, the alternate angles ABC, BCD,
a 29. 1. are equal.[a] Again, because AB is equal to CD, and BC common, the two

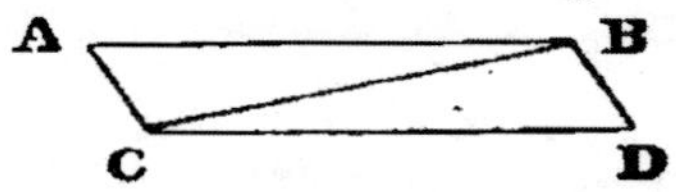

* I think the following mode of enunciating this proposition is clearer than the one in the text. If the extremes of two equal and parallel right lines be joined by the extremes of two other right lines not cutting one another, these two right lines shall also be equal and parallel.

AB, BC, are equal to the two BC, CD, and the angle ABC is equal to the angle BCD. Therefore the base AC is equal to the base BD,[b] and the triangle ABC to the triangle BCD; also the remaining angles will be equal to the remaining angles, each to each, to which the equal sides are opposite. Therefore the angle ACB is equal to the angle CBD. And because the right line BC falls upon the two right lines AC, BD, makes the alternate angles ACB, CBD, equal to one another, AC is parallel to BD.[c] But it was shown that it was equal to it. Therefore right lines, &c. Q. E. D.

[b] 4. 1.

[c] 27. 1.

PROPOSITION XXXIV.

THEOREM.

The opposite sides and angles of a parallelogram are equal to one another, and the diameter bisects it.

"A parallelogram is a four-sided figure, whose opposite sides are parallel."

Let ABDC be a parallelogram, whose diameter is BC; the opposite sides and angles of the parallelogram ABDC are equal to one another, and the diameter BC bisects it. For because AB is parallel to CD, and the right line BC falls upon them, the alternate angles ABC, BCD, are equal to one another.[a] Again, because AC is parallel to BD, and BC falls upon them, the alternate angles ACB, CBD, are equal to one another. Therefore the two triangles ABC, CBD, have the two angles ABC, BCA, equal to the two angles BCD, CBD, each to each, and one side equal to one side, viz. BC, which is common to both. Therefore they will have the remaining sides equal to the remaining sides, each to each, and the remaining angle equal to the remaining angle.[b] Therefore the side AB is equal to the side CD, and the side AC to the side BD; also the angle BAC is equal to the angle BDC. And because the angle ABC is equal to the angle BCD, and the angle CBD to the angle ACB, the whole angle ABD will be equal to the whole angle ACD. But it was shown that the angle BAC is equal to the angle BDC. Therefore

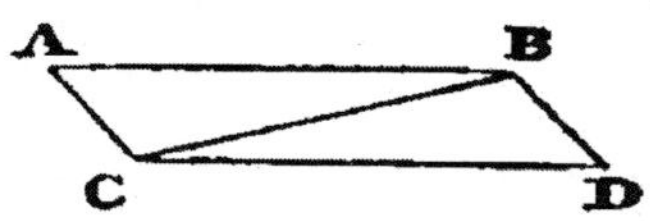

[a] 29. 1.

[b] 26. 1.

the opposite sides and angles of parallelograms are equal to one another, also the diameter bisects it. For because AB is equal to CD, and BC common, the two AB, BC, are equal to the two DC, CB, each to each; and the angle ABC is equal to the angle BCD. There-
c 4. 1. fore the base AC is equal to the base DB,[c] and the triangle ABC is equal to the triangle BCD. Therefore the diameter BC bisects the parallelogram ABCD. Q. E. D.

Deductions.

1. The diameters of a parallelogram bisect each other.
2. If the corresponding diameters of two equiangular parallelograms be equal to one another, also a side of the one equal to the corresponding side of the other, then shall the other opposite sides of the one be equal to the other opposite sides of the other.
3. If in the sides of a square, at equal distances from the four angles, four other points be taken, one in each side, the figure contained by the right lines which join them shall also be a square.

PROPOSITION XXXV.

THEOREM.

Parallelograms constituted upon the same base, and between the same parallels, are equal to one another.

Let ABCD, EBCF, be parallelograms, placed upon the same base BC, and between the same parallels AF, BC. The parallelogram ABCD is equal to the parallelogram EBCF. For because ABCD is a parallelogram, AD is equal to
a 34. 1. BC.[a] For the same reason EF is equal to BC. And therefore AD will be equal to
b Ax. 1. EF[b] and DE common. Therefore the
c Ax. 2. whole AE is equal to the whole DF.[c] But AB is equal to DC. Therefore the two EA, AB, are equal to the two FD, DC, each to each, and the angle EDC is equal
d 29. 1. to the angle EAB, the exterior to the interior;[d] there-
e 4. 1. fore the base EB is equal to the base FC,[e] and the triangle EAB is equal to the triangle FDC. Take away DGE, which is common. Therefore the remaining trapezium ABGD is equal to the remaining trapezium
f Ax. 3. EGCF.[f] Add the triangle GBC, which is common.

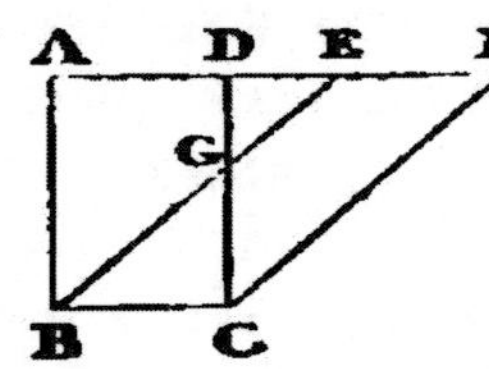

Therefore the whole parallelogram ABCD will be equal to the whole parallelogram EBCF. Therefore parallelograms placed upon the same base, &c. Q. E. D.

PROPOSITION XXXVI.

THEOREM.

Parallelograms constituted upon equal bases, and between the same parallels, are equal to one another.

Let ABCD, EFGH, be parallelograms constituted upon equal bases BC, FG, and between the same parallels AH, BG. The parallelogram ABCD is equal to the parallelogram EFGH. For join BE, CH; and because BC is equal to FG,[a] and FG equal to EH, BC will also be equal to EH. Therefore EB, CH, are both equal and parallel. But those lines are parallel which join the extremities of equal and parallel right lines towards the same parts. Therefore EB, CH, are equal and parallel; wherefore EBCH is a parallelogram, and it is equal to the parallelogram ABCD, for it is placed upon the same base BC, and between the same parallels BC, AD. For the same reason the parallelogram EFGH is equal to the parallelogram EBCH,[b] for it has the same base EH, and is constituted between the same parallels EH, BG. Therefore the parallelogram ABCD will be equal to the parallelogram EFGH. Therefore parallelograms, &c. Q. E. D.

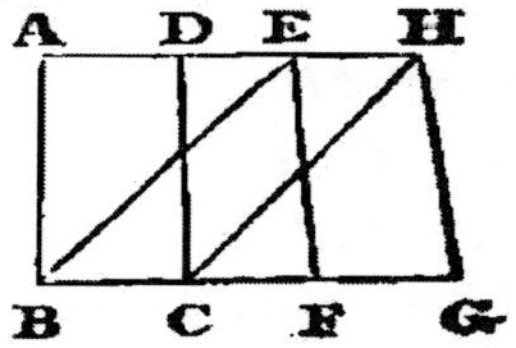

[a] Hyp.

[b] 35. 1.

Deductions.

1. If the sides of a parallelogram be bisected, the lines joining the opposite points of section will divide the parallelogram into four equal parallelograms.

2. If a trapezium have two of its sides parallel to one another, and equal to two sides of another trapezium, which are parallel to one another; also if the perpendicular distance of the one be equal to the perpendicular distance of the other, then shall the two trapeziums be equal to one another.

PROPOSITION XXXVII.

Theorem.

Triangles constituted upon the same base, and between the same parallels, are equal to one another.

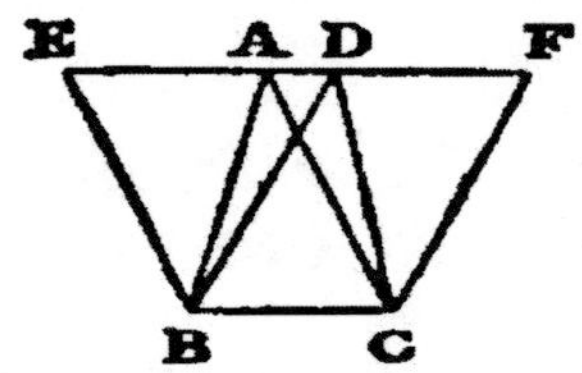

Let the triangles ABC, DBC, be constituted upon the same base BC, and between the same parallels AD, BC. The triangle ABC is equal to the triangle DBC. Produce AD both ways to the points E, F; and through B draw BE,
a 31. 1. parallel to CA,[a] and through C draw CF, parallel to BD. Therefore each of them EBCA, DBCF, is a parallelogram, and the parallelogram EBCA is equal to the parallelogram
b 35. 1. DBCF.[b] For they are upon the same base BC, and between the same parallels BC, EF. And the triangle
c 34. 1. ABC is half of the parallelogram EBCA,[c] because the diameter AB bisects it; and the triangle DBC is half of the parallelogram DBCF, for the diameter DC bisects it. But the halves of equal things are equal. Therefore the triangle ABC is equal to the triangle DBC. Therefore triangles, &c. Q. E. D.

PROPOSITION XXXVIII.

Theorem.

Triangles constituted upon equal bases, and between the same parallels, are equal to one another.

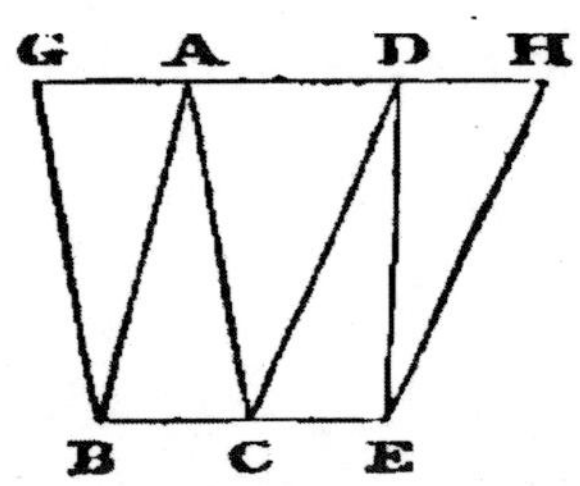

Let the triangles ABC, DCE, be constituted upon equal bases BC, CE, and between the same parallels BE, AD. The triangle ABC is equal to the triangle DCE. For produce AD both ways to the points G, H. Through B draw BG pa-
a 31. 1. rallel to CA;[a] also through E draw EH parallel to DC. Therefore each of the figures GBCA, DCEH, is a parallelogram. And the parallelogram GBCA is equal to the parallelogram
b 36. 1. DCEH,[b] because they are upon equal bases BC, CE, and between the same parallels BE, GH. But the triangle ABC is half of the parallelogram GBCA, for the diameter
c 34. 1. AB bisects it;[c] and the triangle DCE is half of the parallelogram DCEH, for the diameter DE bisects it.
d Ax. 7. But the halves of equal things are equal:[d] therefore

the triangle ABC is equal to the triangle DCE. Therefore triangles, &c. Q. E. D.

Deductions.

1. A right line drawn from the vertex of a triangle bisecting the base, divides the triangle into two equal triangles.

2. If two opposite sides of a trapezium be parallel to one another, the line joining their bisections bisects the trapezium.

PROPOSITION XXXIX.

THEOREM.

Equal triangles placed upon the same base, and towards the same parts, are also between the same parallels.

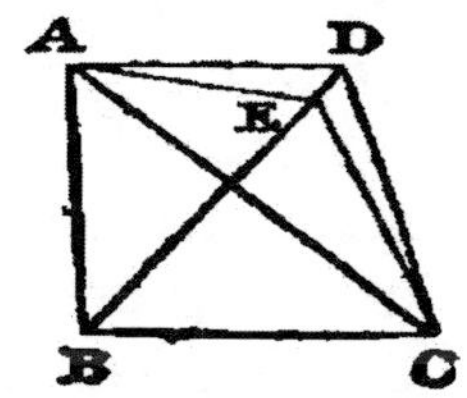

Let the equal triangles ABC, DBC, be constituted upon the same base BC, and towards the same parts they are between the same parallels. For draw AD; AD is parallel to BC. For if it is not parallel, draw through the point A the right line AE, parallel to BC,[a] and join EC. Therefore the triangle ABC is equal to the triangle EBC, because it is upon the same base BC, and between the same parallels BC, AE.[b] But the triangle ABC is equal to the triangle DBC. Therefore also, the triangle DBC is equal to the triangle EBC, the greater to the less, which is impossible. Therefore AE is not parallel to BC. In like manner we show that none other, except AD, is parallel to BC. Whence AD is parallel to BC. Therefore equal triangles, &c. Q. E. D.

[a] 31. 1.
[b] 37. 1.

PROPOSITION XL.

THEOREM.

Equal triangles constituted upon equal bases, and towards the same parts, are also between the same parallels.

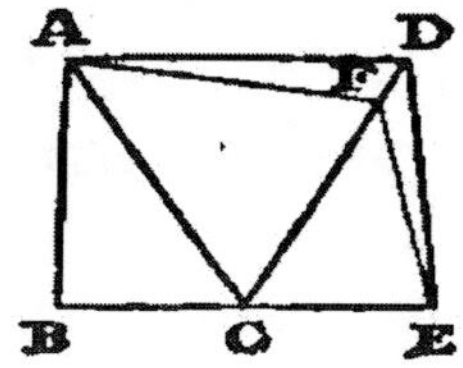

Let the equal triangles ABC, CDE, be constituted on the equal bases BC, CE. They are between the same parallels. Draw AD; AD is parallel to BE. For if it is not, through A draw AF parallel to BE,[a] and join DE. Therefore the triangle ABC is equal to the triangle CEF,[b] because

[a] 31. 1.
[b] 38. 1.

they are constituted between the same parallels BE, AF, and upon equal bases. But the triangle ABC is equal to the triangle DCE. Therefore also the triangle DCE will be equal to the triangle FCE, the greater to the less, which is impossible. Therefore AF is not parallel to BE. In like manner we may show that no other line drawn through A is parallel to BE except AD. Therefore AD will be parallel to BE. Therefore equal triangles, &c. Q. E. D.

PROPOSITION XLI.

THEOREM.*

If a parallelogram and a triangle have the same base, and are between the same parallels, the parallelogram will be double of the triangle.

For let ABCD be a parallelogram, and EBC a triangle; let them have the same base BC, and between the same parallels BC, AE. The parallelogram ABCD, is double of the triangle EBC. For join AC. Therefore the triangle

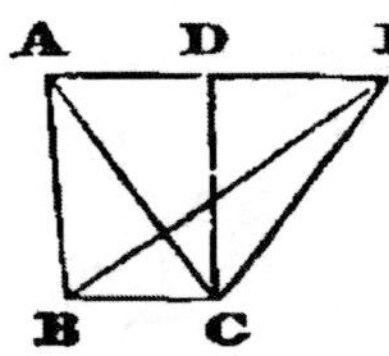

a 37. 1. ABC is equal to the triangle EBC,[a] for they are constituted upon the same base BC, and between the same parallels BC, AE. But the parallelogram ABCD is double of the triangle ABC, because the

b 34. 1. diameter AC bisects it.[b] Wherefore it will also be double of the triangle EBC. If therefore a parallelogram and a triangle, &c. Q. E. D.

PROPOSITION XLII.

PROBLEM.

To make a parallelogram equal to a given triangle, and having one of its angles equal to a given rectilineal angle.

Let ABC be a given triangle, and D a given rectilineal angle. It is required to make a parallelogram equal to the triangle ABC, and having one of its angles equal to the given recti-

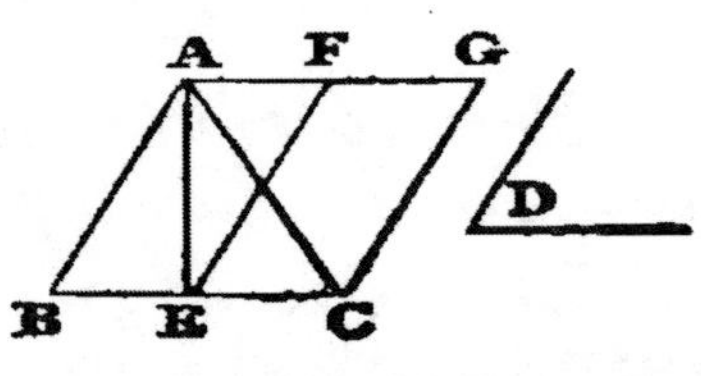

a 10. 1. lineal angle D. Bisect BC in E;[a] and AE being joined to the right line EC and to the point in it E, make the

* From this proposition is derived the rule for finding the area of a triangle, the base and altitude being given; for as the area of a parallelogram is the product of the base and altitude, it follows that the area of a triangle must be half that product.

angle CEF equal to D;[b] and through the point A draw [b] 23. 1.
AG parallel to EC;[c] and through C draw CG parallel to [c] 31. 1.
FE; therefore FECG is a parallelogram. And because BE is equal to EC, the triangle ABE will be equal to
the triangle AEC,[d] because they are upon equal bases [d] 38. 1.
BE, EC, and between the same parallels, BC, AG. Therefore the triangle ABC is double of the triangle AEC. But the parallelogram FECG is also double of the triangle
AEC,[e] because it has the same base, and is between the [e] 41. 1.
same parallels. Therefore the parallelogram FECG is equal to the triangle ABC, and hath an angle CEF equal to the given angle D. Therefore a parallelogram FECG has been made equal to the triangle ABC, and having an angle CEF equal to the angle D.* Q. E. F.

Deductions.

1. A trapezium is equal to half the rectangle, whose base is the diagonal of the trapezium, and perpendicular the aggregate of the perpendiculars drawn from the vertical angles unto the diagonal.

2. To describe a parallelogram, the area and perimeter of which shall be respectively equal to the area and perimeter of a given triangle.

3. If the sides of a parallelogram be bisected, the lines joining the points of bisection shall contain a parallelogram equal to half of the given one, and the diameters of this parallelogram shall be equal to half of the perimeter of the other.

PROPOSITION XLIII.

THEOREM.

The complements of any parallelogram which are about the diameter of any parallelogram are equal to one another.

Let ABCD be a parallelogram whose diameter is BD. Let FH, EG, be the parallelograms, about the diameter

* Hence also it may be shown, that triangles which are equal, and between the same parallels, are either upon the same base, or upon equal bases: thus let there be two triangles ABC, DEF, which are equal, and between the same parallels AD, BF, the bases BC, EF, are also equal. For if they are not, let BC be the greater, and from it cut off BH equal to EF, and join AH. Therefore because the triangles ABH, DEF, are upon equal bases BH, EF, and between the same parallels AD, BF, they are equal. But the triangles ABC, DEF, are equal. Whence the triangle ABC is equal to the triangle ABH, the greater to the less, which is impossible. Therefore the base BC is not unequal to the base EF; that is, it is equal to it. And this method of demonstration is the same in parallelograms.

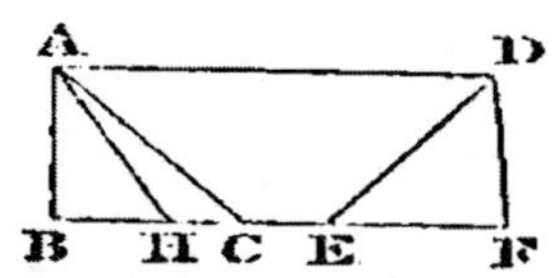

BD, and AK, KC, the complements. The complement AK is equal to the complement KC. And because ABCD is a parallelogram, and BD its diameter, the triangle ABD is equal to the triangle BCD.[a] Again because HKFD is a parallelogram whose diameter is DK, the triangle HDK will be equal to the triangle DKF. For the same reason the triangle KGB is equal to the triangle KEB. Therefore the triangle BEK is equal to the triangle BGK, and the triangle HDK to the triangle DKF; and the triangle BEK, together with the triangle HDK, will be equal to the triangle BGK, together with the triangle DKF. But the whole triangle ABD is equal to the whole BDC. Therefore the remaining complement AK is equal to the remaining complement KC. Therefore the complements, &c. Q. E. D.

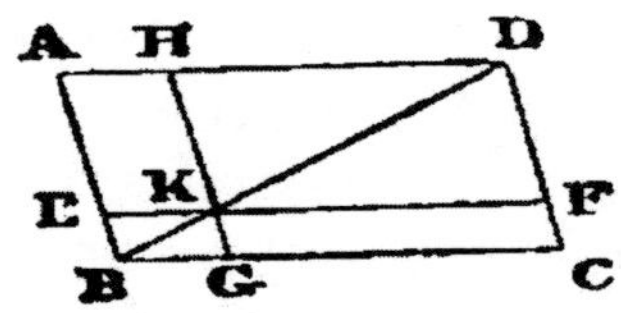

[a] 34. 1.

PROPOSITION XLIV.

Problem.*

To a given right line to apply a parallelogram equal to a given triangle, and having an angle equal to a given rectilineal angle.

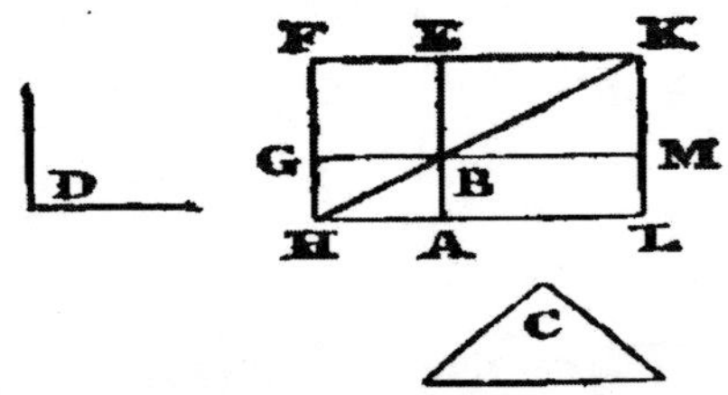

Let AB be the given right line, C the given triangle, and D the given rectilineal angle. It is required to the right line AB to apply a parallelogram equal to the given triangle C, and having an angle equal to D. Describe a parallelogram BEFG equal to the triangle C,[a] and having an angle EBG equal to the angle D; and put BE in a direct line with AB produce FG to H; and through A draw AH parallel to BG or EF,[b] and join HB. And because the right line HF falling upon the parallels AH, EF, the angles AHF, HFE, are equal to two right angles.[c] Wherefore BHF, HFE, are less than two right angles. But those right lines, which with another falling upon them make the adjacent angles less than two right

[a] 42. 1.

[b] 31. 1.

[c] 29. 1.

* Edmund Stone thus enunciates this proposition; make such a parallelogram that a given right line shall be one of its sides; one of its angles shall be equal to a given right lined angle, and the parallelogram shall be equal to a given triangle.

angles, do meet if produced far enough.[d] Therefore [d] Post. 5.
HB, FE, being produced, meet. Let them be produced and meet in K; and through K draw KL parallel either to EA or FH, and produce AH, GB, to the points LM. Therefore HLKF is a parallelogram whose diameter is HK; and AG, ME, are the parallelograms about HK, and LB, BF, are those called complements. Therefore LB is equal to BF.[e] But BF is equal to the triangle C. [e] 43. 1.
Wherefore also LB will be equal to the triangle C. And because the angle GBE is equal to the angle ABM,[f] but [f] 15. 1.
it is also equal to the angle D, therefore the angle ABM will also be equal to the angle D. Therefore to a given right line AB a parallelogram has been applied equal to the given triangle C, and having an angle equal to the given rectilineal angle D. Q. E. F.

PROPOSITION XLV.

PROBLEM.

To describe a parallelogram equal to a given rectilineal figure, and having an angle equal to a given rectilineal angle.

Let ABCD be the given rectilineal figure, and E the given rectilineal angle: it is required to describe a parallelogram equal to the given rectilineal figure ABCD, and having an angle equal to E. Join DB, and describe the parallelogram FH equal to the triangle ADB, and having the angle HKF equal to the angle E.[a] [a] 42. 1.

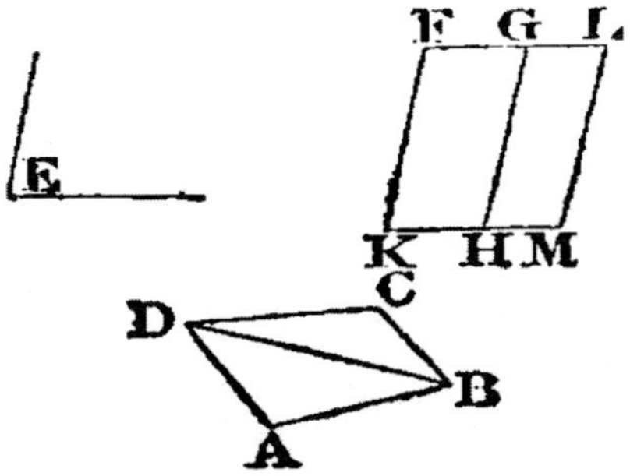

Then to the right line GH apply the parallelogram GM equal to the triangle DBC, and having the angle GHM equal to the angle E.[b] And because the angle E is [b] 44. 1.
equal to each of the angles HKF, GHM, the angle GHM will also be equal to the angle HKF; add KHG, which is common; the angles HKF, KHG, are equal to the angles KHG, GHM. But HKF, KHG, are equal to two right angles.[c] [c] 29. 1.
Therefore KHG, GHM, will be equal to two right angles. Therefore to the right line GH, and to the point H in it, two right lines KH, HM, not placed towards the same parts, make the adjacent angles equal to two right angles: KH is in a direct line with HM.[d] [d] 14. 1.
And because the right line HG falls upon the parallels KM, FG, the alternate angles MHG, HGF, are equal to

one another; add HGL, which is common: therefore the angles MHG, HGL, are equal to the angles HGF, HGL. But the angles MHG, HGL, are equal to two right angles. Wherefore the angles HGF, HGL, will also be equal to two right angles. Therefore FG is in a right line with GL. And because KF is parallel and equal to HG, HG
[c] 30. 1. is also equal and parallel to ML;[c] therefore likewise KF will be equal and parallel to ML. Join the right lines KM, FL; therefore KM, FL, are both equal and parallel. Therefore KFLM is a parallelogram. But the triangle ABD is equal to the parallelogram FH; also the triangle DBC equal to the parallelogram GM. The whole parallelogram KLFM is equal to the whole rectilineal figure ABCD. Therefore the parallelogram KFLM is equal to the given rectilineal figure ABCD, and having the angle FKM equal to the given angle E. Q. E. F.

Deduction.

To a given right line to apply a parallelogram equal to a given rectilineal figure, and having an angle equal to a given rectilineal angle.

PROPOSITION XLVI.

Problem.*

Upon a given right line to describe a square.

Let AB be the given right line. It is required upon the line AB to describe a square. Draw from the given
[a] 11. 1. point A the right line AC, at right angles to AB,[a] and
[b] 3. 1. make AD equal to AB,[b] and through the point D draw DE parallel to AB; also
[c] 31. 1. through B draw BE parallel to AD.[c] Therefore ADEB is a parallelogram. And AB is
[d] 34. 1. equal to DE;[d] also AD to EB, but BA is equal to AD. Therefore the four, BA, AD, DE, EB, are equal to one another. Therefore the parallelogram ADEB is equilateral; it is also rectangular. For because the right line AD falling upon the parallels AB, DE, the angles BAD, ADE, are
[e] 29. 1. equal to two right angles;[e] but BAD is a right angle; therefore ADE will also be a right angle; but the opposite sides and angles of parallelograms are equal to one

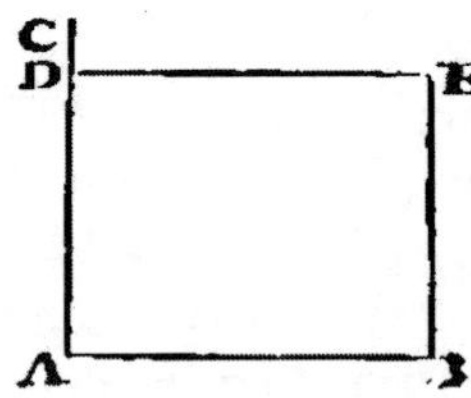

* In like manner a rectangle may be described which is contained under two right lines.

another. Therefore each of the opposite angles ABE, BED, is a right angle; and consequently ABDE is a rectangle. But it was shown to be equilateral, therefore it is a square, and has been described on the given right line AB. Q. E. F.

COROLLARY.

Hence every parallelogram having one right angle is a rectangle.

PROPOSITION XLVII.

THEOREM.

In right angled triangles the square described upon the side subtending the right angle is equal to the squares described on the sides containing the right angle.

Let ABC be a right angled triangle, having the right angle BAC. The square described upon the right line BC is equal to the squares described on the sides BA, CA. Describe on BC the square BDEC,[a] and on BA, AC, the squares GB, HC; and through A draw AL parallel either to BD or CE, and join AD, FC. Therefore because the angles BAC, BAG, are each of them right angles, the two right lines AC, AG,

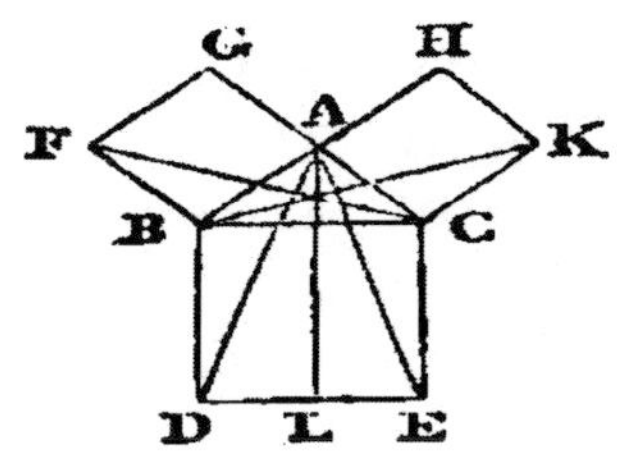

a 46. 1.

upon the opposite sides of AB, make with it, at the point A, the adjacent angles equal to two right angles:
GA, AC, are in one and the same right line.[b] For the b 14. 1.
same reason AB, AH, are in one and the same right line. And because the angle DBC is equal to the angle FBA, each of them is a right angle: add ABC, which is common: therefore the whole angle DBA is
equal to the whole angle FBC.[c] But the two sides c Ax. 2.
AB, BD, are equal to the two sides FB, BC, each to each, and the angle DBA is equal to the angle FBC;
also the base AD will be equal to the base FC;[d] and the d 4. 1.
triangle ABD to the triangle FBC. And the parallelogram BL is double of the triangle ABD, because they have the same base BD, and are between the same
parallels BD, AL;[e] also the square GB is double of the e 41. 1.
triangle FBC; because they have the same base FB, and are between the same parallels FB, GC. But things which are double of equals are equal to one another; therefore the parallelogram BL is equal to the square BG. In like manner AE, BK, being joined, the paral-

lelogram CL may be also shown to be equal to the square HC. Therefore the whole square BDEC, is equal to the two squares GB, HC, and BDEC is the square described upon the side BC, and GB, HC, are the squares described on BA, AC. Therefore the square described on the side BC, subtending the right angle, is equal to the squares described on the sides BA, AC, containing the right angle. Therefore in right angled triangles, &c. Q. E. D.

Deductions.

1. Describe a square which shall be equal to two given squares.
2. Describe a square which shall be equal to the difference between two given squares.
3. The square described upon the diameter of another square is double of that square.
4. If the sides of the square described upon the hypothenuse of a right angled triangle be produced to meet the sides of the squares described upon the legs, they will cut off triangles equiangular, and equal to the given triangle.

The name of Pythagoras is rendered immortal in the annals of geometry by the discovery of this famous, useful, and elegant proposition. Some authors relate that he was so transported with joy, that he offered to the gods a sacrifice of a hundred oxen, as a token of gratitude for their inspiring him with it. This circumstance, however, is doubted by others, as being inconsistent with his religious opinions, which prohibited bloody sacrifices. Be this as it may, never had enthusiasm a better foundation. The problem deservedly ranks amongst the first class of geometrical truths, both from the singularity of its result, and the variety of cases to which it is applicable in every branch of the mathematics.

PROPOSITION XLVIII.

Theorem.

If the square which is described upon one of the sides of a triangle be equal to the squares which are described on the other two sides, then shall the angle which is contained by these two remaining sides be a right angle.

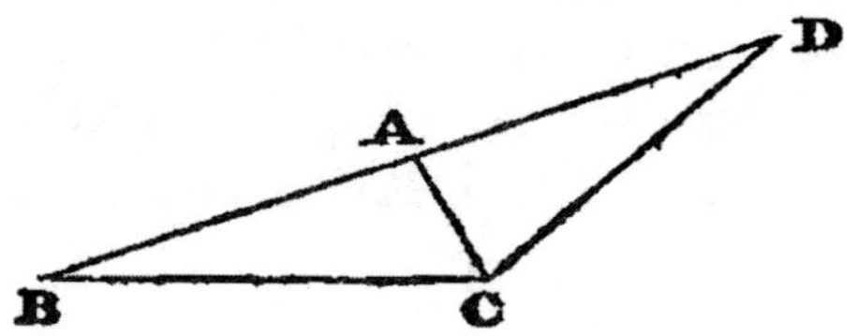

If the square described on BC, one of the sides of the triangle ABC, be equal to the squares described upon the remaining sides of the triangle

BA, AC, the angle BAC is a right angle. For draw from the point A, AD at right angles to AC, and make AD equal to BA, and join DC. Therefore because DA is equal to AB, the square also described on DA is equal to the square described on AB. Add the square of AC, which is common. Therefore the squares of DA, AC, are equal to the squares of BA, AC. But the square described on DC is equal to the squares described on DA, AC, for DAC is a right angle;[a] also the square of BC is equal to the squares of BA, AC. Therefore the square of DC is equal to the square BC. Wherefore the side DC is also equal to the side CB. And because DA is equal to AB, and AC common, the two DA, AC, are equal to the two BA, AC; and the base DC is equal to the base CB. Therefore the angle DAC is equal to the angle BAC.[b] But DAC is a right angle; therefore BAC will also be a right angle. If therefore the square, &c. Q. E. D.

[a] 47. 1.

[b] 8. 1.

EUCLID'S ELEMENTS.

BOOK II.

DEFINITIONS.

1. Every right angled parallelogram is said to be contained under two right lines, comprehending a right angle.

2. In every parallelogram either of those parallelograms about the diameter, together with the complements, is called a gnomon.* Thus the parallelogram HG, together with the complements AF, FC, is the gnomon, which is more briefly expressed by the letters AGK, or EHC, which are at the opposite angles of the parallelograms, which make the gnomon.

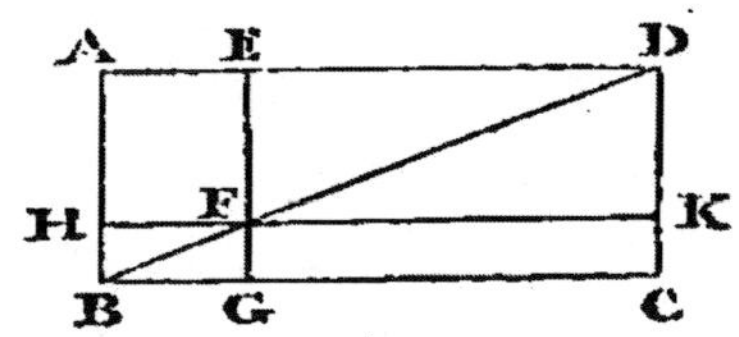

PROPOSITION I.

THEOREM.

If there be two right lines, one of which is divided into any number of parts, the rectangle comprehended under the whole and divided line, is equal to the rectangles contained under the whole line, and the several segments of the divided line.

Let A and BC be two right lines, and let BC be any how divided in the points D, E; the rectangle contained under the right lines A and BC, is equal to the rectangles contained under A and BD, A and DE, and A and EC.

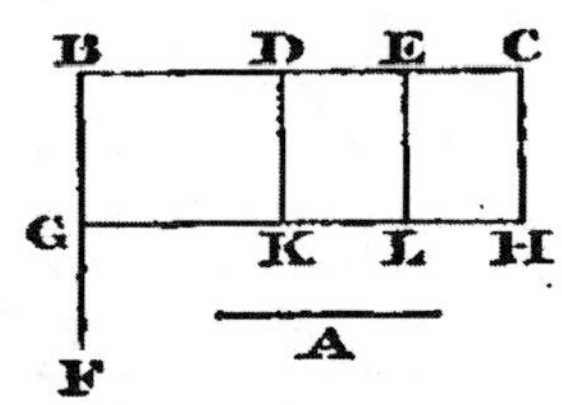

From the point B draw[a] BF, at right angles to BC, and make[b] BG equal to A: and let[c] GH be drawn through G, parallel to BC; and through D, E, C, draw DK, EL, and CH, parallel to BG; then the rectangle BH is equal to the rectangles BK, DL, EH, but the rectangle BH is contained under A and BC, for it is contained under GB and BC,

[a] 11. 1.
[b] 3. 1.
[c] 31. 1.

* From the Greek word Γνωμων, which signifies a carpenter's square.

and GB is equal to A; and BK is contained under A and BD, for it is contained under BG and BD, and BG is equal to A; and the rectangle DL is contained under
d 34. 1. A and DE; because DK, that is[d] BG, is equal to A; so likewise the rectangle EH is that contained under A and EC. Therefore the rectangle under A and BC is equal to the rectangles under A and BD, A and DE, and A and EC. Therefore if there be two right lines, &c. Q. E. D.

The same by Algebra.

Put a equal to the line A, and b the line BC, and suppose it be divided into the parts e, f, g; then shall
* Ax. 8. 1. $ab = ae + af + ag$; for $b^* = e + f + g$, multiply each by a, and we shall have $ab = ae + af + ag$. Q. E. D.†

PROPOSITION II.

Theorem.

If a right line be divided into any two parts, the rectangles contained under the whole and each of the parts are together equal to the square of the whole line.

Let the right line AB be any how divided in the point C, the rectangle contained under AB, BC, together with the rectangle AB, AC, is equal to the square of AB. For let the square ADEB, be de-
a 46. 1. scribed[a] on AB, and through C let CF be
b 31. 1. drawn parallel[b] to AD or BE. Then AE is
c Ax. 8. 1. equal to the rectangles AF, CE;[c] and AE is the square of AB; and AF is the rectangle

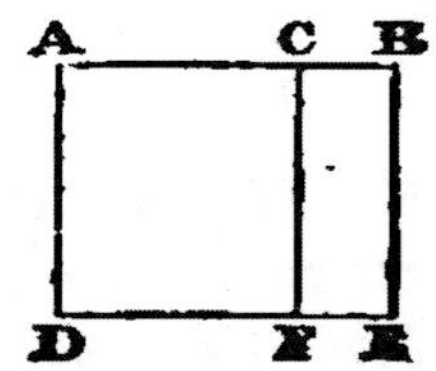

† If two given right lines are both divided into how many parts soever, one whole multiplied into the other shall bring out the same product as the parts of the one multiplied into the parts of the other.

For, let $x = a + b + c$, and $y = d + e$; then because $dx = ad + bd + cd$, and $ex = ae + be + ce$, and $xy = dx + ex$, therefore xy will be $= ad + bd + cd + ae + be + ce$. Q. E. D.

From hence we deduce a method of multiplying compound lines into compound lines. For if the rectangles of all the parts be taken, their sum shall be equal to the rectangle of the wholes.

But whensoever in the multiplication of lines into themselves you meet with the sign + intermingled with —, particular regard must be paid to them. For of + multiplied into — arises —; but of — into — arises +. For example, let $+ a$ be multiplied into $b - c$; then because $+ a$ is not affirmed of all b, but only of that part of it whereby it exceeds c, therefore ac must remain denied; so that the product will be $ab - ac$.

This being sufficiently understood, the nine following propositions, and innumerable others of that kind, arising from the comparing of lines multiplied into themselves, which the reader may find done in Vieta and other analytical writers, are demonstrated with great facility, by reducing the matter for the most part to almost a simple work.

contained under AB, AC; for it is contained under AD, AC, whereof AD is equal to AB; and the rectangle CE is contained under AB, BC, since BE is equal to AB. Therefore the rectangle under AB and AC, together with the rectangle under AB and BC, is equal to the square of AB. If therefore a right line, &c. Q. E. D.

Deduction.

If a right line be divided into any number of parts, the rectangles contained by the whole and each of the parts are together equal to the square of the whole line.

The same by Algebra.

Let a equal the line AB, and suppose it divided into any two parts e, f; then shall $ae + af = a^2$; for a* $= e$ * Ax. 8, 1.
$+ f$; multiply by a, and we shall have $ae + af = a^2$. Q. E. D.

PROPOSITION III.

THEOREM.

*If a right line be any how divided into two parts, the rectangle contained under the whole line and one of its parts is equal to the rectangle contained under the two parts, together with the square of the aforesaid part.**

Let the right line AB be divided into any two parts in the point C; the rectangle AB, BC, is equal to the rectangle AC, CB, together with the square of BC. For on BC describe[a] the square BCDE, and let ED be produced to F, and through A let AF be drawn[b] parallel to CD or BE; then the rectangle AE is equal to the rectangles AD, CE, and AE is the rectangle contained under AB, BC, for it is contained under AB, BE, whereof BE is equal to BC, and AD is contained under AC, CB, for CD is equal to CB; and DB is the square of BC; therefore the rectangle under AB, BC, is equal to the rectangle under AC, CB, together with the square of BC. If therefore a right line, &c. Q. E. D.

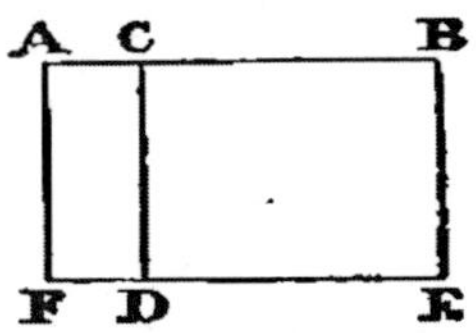

a 46. 1.
b 31. 1.

The same by Algebra.

Put a equal to the right line AB, and suppose it

---

* In the translations of Commandine and Gregory this proposition appears ambiguously enunciated, as no mention is made of the number of parts into which the right line should be cut.

divided into any two parts f, g,* then is $af = f^2 + gf$,
† Ax. 8. 1. or $ag = fg + g^2$. For a† $= f + g$; multiply by f, and we shall have $af = f^2 + fg$; or multiply by g, and it will be $ag = fg + g^2$. Q. E. D.

PROPOSITION IV.

Theorem.

If a right line be divided into any two parts, the square of the whole line is equal to the squares of the two parts, together with twice the rectangle contained under the parts.

Let the right line AB be divided into any two parts in the point C; the square of AB is equal to the squares of AC, CB, and to twice the rectangle con-
a 46. 1. tained under AC, CB. Upon AB describe[a] the square ADEB, and join BD, and through
b 31. 1. C draw[b] CGF parallel to AD or BE, and through G draw HK parallel to AB or DE, and because CF is parallel to AD, and BD

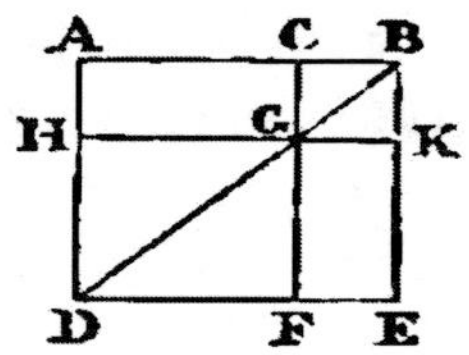

c 29. 1. falls upon them, the exterior angle BGC is equal[c] to the
d 5. 1. interior and opposite angle ADB, but ADB is equal[d] to the angle ABD, because BA is equal to AD, being sides of a square; wherefore the angle GCB is equal to the angle GBC: and therefore the side BC is equal to the side
e 6. 1. CG.[e] But also the side CB is equal[f] to GK, and CG to
f 34. 1. BK; wherefore the figure CGKB is equilateral; it is also rectangular;* for the angle CBK is a right one, but a
g Cor. 46. 1. parallelogram, having one right angle, is right angled;[g] wherefore CGKB is a rectangle; but it has also been proved to be equilateral. Wherefore CGKB is a square described upon BC. For the same reason HF is also a square made upon HG, that is, equal to the square of AC. Wherefore HF and CK are the squares of AC and CB, and
h 43. 1. because the rectangle AG is equal[h] to the rectangle GE, and AG is that contained under AC and CB, for GC is equal to CB: GE shall be equal to the rectangle under AC, CB; wherefore the rectangles AG, GE, are equal to twice the rectangle contained under AC, CB; and HF, CK, are the squares of AC, CB. Wherefore the four figures HF, CK, AG, GE, are equal to the squares of AC, CB, and to twice the rectangle AC, CB. But HF, CK, AG, GE, make up the whole figure ADEB, which is the

* Dr. Simson, and others, in proving the figure CGKB to be rectangular, gives a long demonstration; whereas it is easily deduced from the 46th prop. 1st Book, that any parallelogram having one right angle is a rectangle.

square of AB. Therefore the square of AB is equal to the squares of AC, CB, together with twice the rectangle contained under AC, CB. Wherefore if a right line, &c. Q. E. D.

Deductions.

1. The parallelograms which stand about the diameter of a square, are likewise squares.
2. The diameter of any square bisects its angles.
3. If a line be divided into two equal parts, the square of the whole line will be equal to four times the square of half the line.

The same by Algebra.

Put a equal to the right line AB, and suppose it divided into any two parts f, g; then shall $a^2 = f^2 + 2fg + g^2$. For a[*] $= f + g$ square each side, and we shall have $a^2 = f^2 + 2fg + g^2$. Q. E. D. [*] Ax. 8. 1.

Otherwise.

$af = f^2 + fg$,[†] and $ag = fg + g^2$; whereas $af + ag$[‡] $= a^2$, thence is $a^2 = f^2 + 2fg + g^2$. Q. E. D. [†] 3. 2. [‡] 2. 2.

PROPOSITION V.

THEOREM.

If a right line be divided into two equal parts, and two unequal ones, the rectangle under the unequal parts, together with the square that is made of the intermediate distance, is equal to the square made of half the line.

Let the right line AB be divided into two equal parts in the point C, and into two unequal parts at the point D; the rectangle AD, DB, together with the square of CD, is equal to the square CB.

For describe[a] CEFD, the square of CB, and join BE, draw[b] DHG through D, parallel to CE or BF, and KLO through H, parallel to CB or EF, as also AK through A parallel to CL, or BO. [a] 46. 1. [b] 31. 1.

Now the complement CH is equal[c] to the complement HF; to each of these add DO, and the whole CO is equal to the whole DF, but CO is equal to AL, because AC is equal to CB; therefore AL is equal to DF, and adding [c] 43. 1.

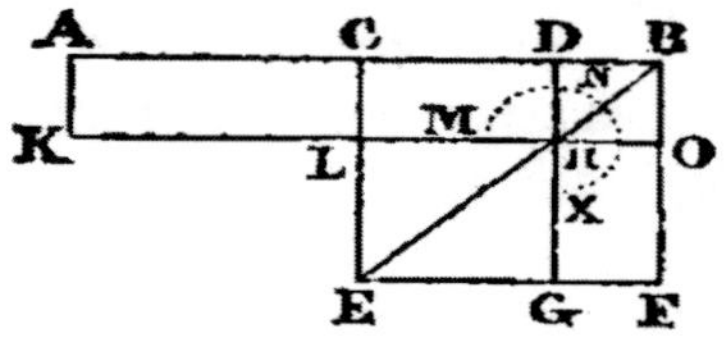

CH, which is common, the whole AH shall be equal to FD, DL, together. But AH is the rectangle contained under AD, DB; for DH is equal[d] to DB, and FD, DL, is the gnomon MNX; therefore MNX is equal to the rectangle under AD, DB; and if LG, being common, and equal[d] to the square of CD be added; therefore the gnomon MNX and LG are equal to the rectangle under AD, DB, together with the square of CD; but the gnomon MNX and LG make up the whole square CEFB, i. e. the square of CB. Wherefore the rectangle under AD, DB, together with the square of CD, is equal to the square of CB. If, therefore, a right line, &c.

[d] Cor. 4. 2.

Deduction.

If a right line be divided into two unequal parts in two different points, the rectangle contained by the two parts which are the greatest and the least, is less than the rectangle contained by the other two parts; the squares of the two former parts together are greater than the squares of the two latter taken together; and the difference between the squares of the former and the squares of the latter is the double of the difference between the two rectangles.

The same by Algebra.

Put a equal to the line AB, e = AD, and f = DB.

$\therefore f = a - e$*

$fe = ae - e^2$, by multiplying by e.

$\therefore fe + e^2 - ae = 0$.

$\therefore fe + \overline{e - \frac{a}{2}}^2 = \frac{a^2}{4}$

or $AD \cdot BD + CD^2 = BC^2$. Q. E. D.

* Ax. 8. 1.

PROPOSITION VI.

THEOREM.

If a right line be divided into two equal parts, and produced to any point, the rectangle contained under the whole line thus produced, and the part produced, together with the square of half the line, is equal to the square of the right line, which is made up of the half and the part produced.

Let the right line AB be bisected in the point C, and produce it to the point D; the rectangle AD, DB, together with the square of CB, is equal to the square of CD.

For on CD describe[a] the square CEFD, and join DE; [a] 46. 1.
through B draw[b] BHG parallel to either CE or DF; and [b] 31. 1.
through H draw KLM parallel to AD or EF, as also AK through A parallel to CL or DM. Therefore because AC is equal to CB, the rectangle AL shall be equal[c] to the [c] 36. 1.
rectangle CH; but CH is equal[d] to HF; and therefore [d] 43. 1.
AL shall be equal to HF; and adding CM, which is common to both, then the whole rectangle AM is equal to the gnomon NXO.

But AM is the rectangle contained under AD, DB, for DM is equal[e] to DB; therefore the gnomon NXO is equal [e] Cor. 4. 2.
to the rectangle under AD, DB, and adding LG, which is common, i. e. the square of CB; and then the rectangle under AD, DB, together with the square of BC, is equal to the gnomon NXO, and LG. But the gnomon NXO, and LG, together, make up the figure CEFD; that is, the square of CD. Therefore the rectangle under AD, and DB, together with the square of BC, is equal to the square of CD. Therefore, if a right line, &c. Q. E. D.

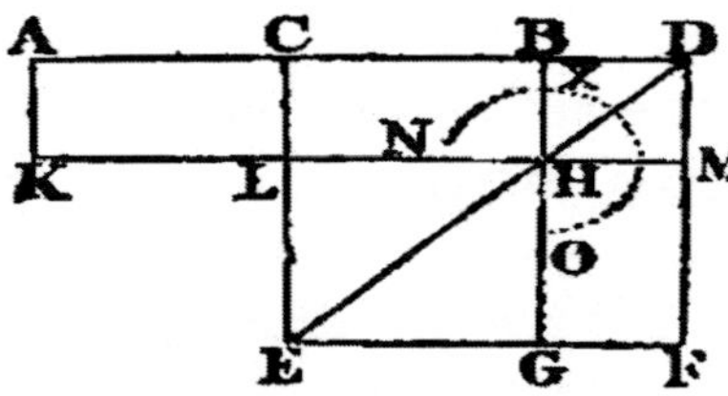

The same by Algebra.

Put a equal to the line AB, and e equal to the added line BD, then shall $ae + e^2 + \frac{1}{4}a^2 = \overline{\frac{1}{2}a + e}^2$; whereof $ae + e^2$ is the rectangle under AD, DB; $\frac{1}{4}a^2$ the square of CB and $\overline{\frac{1}{2}a + e}^2$ the square of CD. For $ae + e^2 + \frac{1}{4}a^2 = \overline{\frac{1}{2}a + e}^2 = ae + e^2 + \frac{1}{4}a^2$. Q. E. D.

Deduction.

If three right lines $a, a + \frac{1}{2} b, a + b$, be in arithmetical proportion, then the rectangle contained under the extreme terms $a, a + b$, together with the square of the difference $\frac{1}{2} b$, is equal to the square of the middle term $a + \frac{1}{2} b$.

PROPOSITION VII.

Theorem.

If a right line be any how divided into two parts, the square of the whole line together with the square of one of the parts, is equal to double the rectangle contained under the whole line, and the said part, together with the square made of the other part.

Let the right line AB be any how cut in the point C; the squares of AB, BC, are equal to twice the rectangle under AB, BC, together with the square of AC.

[a] 46. 1. For let the square ADEB be described[a] on AB, and construct the figure as in the preceding propositions;

[b] 43. 1. and because AK is equal[b] to KE, add to each of them CF; the whole AF is therefore equal to the whole CE; therefore AF, CE, are double of AF. But AF, CE, are the gnomon KLM, together with the square CF; therefore the gnomon KLM, and the square CF, will be double of the rectangle AF, or double of the

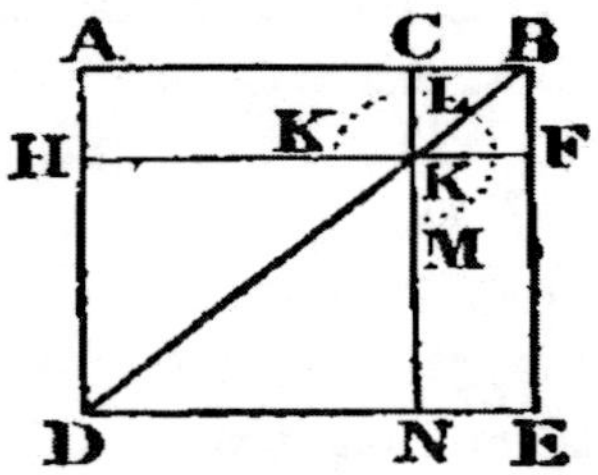

[c] Cor. 4. 2. rectangle under AB, BC; for BF is equal[c] to BC. To each of these equals, add HN, which is the square of AC; then the gnomon KLM, and the squares CF, HN, are equal to double the rectangle contained under AB, BC, with the square of AC. But the gnomon KLM, together with the squares CF, HN, are equal to ADEB, and CF, which are the squares of AB, BC. Therefore the squares of AB, BC, are together double of the rectangle under AB, BC, together with the square of AC. Therefore, if a right line, &c. Q. E. D.

Deductions.

1. The square of the difference of any two lines is equal to the square of both the lines less by a double rectangle comprehended under the said lines.

2. The sum of the squares of the sum and difference of two lines is equal to twice the sum of the squares of those lines, and the difference of the squares of the sum and difference of two lines is equal to four times the rectangle contained by those lines.

3. The squares of any two unequal right lines are together greater than twice the rectangle contained by those lines.

The same by Algebra.

Let a be put equal to the line AB, and suppose it divided into any two parts e, f; then shall $a^2 + e^2 = 2\,a\,e + f^2$; whereof a^2 denotes the square of AB; e^2 the square of one of its parts, viz. CB; $2\,a\,e$, double the rectangle under AB, BC; and f^2, the square of AC. For $a = e + f$, whence $a^{2*} = e^2 + 2\,ef + f^2$, and $2\,a\,e\dagger = 2\,e^2 + 2\,ef$; add this to the preceding equation, and it will be $a^2 + 2\,e^2 + 2\,ef\ddagger = 2\,a\,e + e^2 + 2\,ef + f^2$; subtract $e^2 + 2\,ef$ from each quantity, and we shall have $a^2 + e^2\S = 2\,a\,e + f^2$. Q. E. D. * 4. 2. † 3. 2. ‡ Ax. 2. 1. § Ax. 3. 1.

PROPOSITION VIII.

THEOREM.

If a right line be any how cut into two parts, four times the rectangle, contained under the whole line, and one of the parts, together with the square of the other part, is equal to the square of the line, compounded of the whole line, and the first part taken as one line.

Let the right line AB be divided into any two parts in the point C; four times the rectangle contained under AB, BC, together with the square of AC, is equal to the square of the right line made up of AB and BC together.

For let the right line AB be produced to D, so that BD is equal to BC, describe the square AEFD on AD, and construct the double figure (as in the preceding propositions).

Now since CB is[a] equal to BD, and also to GK,[b] and BD is equal to KN; GK shall be likewise equal to KN; by the same reasoning, PR is equal to RO. And since CB is equal to BD, and GK to KN, the rectangle CK will[c] be equal to the rectangle BN, and the rectangle GR to the rectangle RN. But CK is equal to RN, [a] Hyp. [b] 34. 1. [c] 36. 1.

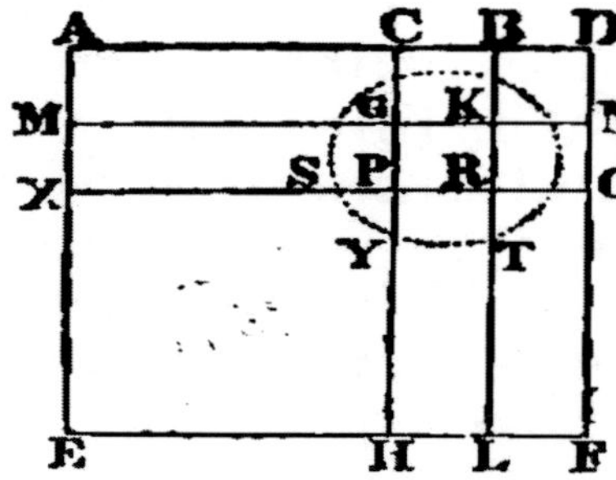

because they are the complements of the parallelogram CO; therefore also BN is equal to GR; and the four rectangles BN, CK, GR, RN, are therefore equal to another, and so are quadruple of the rectangle CK. Again, because CB is equal to BD, and BD to BK; that is, to CG, and
[d] Cor. 4. 2. CB equal to GK; that is[d] to GP, therefore CG is equal to GP; and because CG is equal to GP, and PR to RO, the rectangle AG is equal to the
[e] 43. 1. rectangle MP, and PL to RF. But MP is equal[e] to PL; for they are the complements of the parallelogram ML; wherefore AG is equal to RF. Therefore the four rectangles AG, MP, PL, RF, are equal among themselves, and so are quadruple of one of them AG. And it was demonstrated that the four CK, BN, GR, RN, are quadruple of CK. Therefore the eight rectangles containing the gnomon STY are quadruple of AK. And because AK is that contained under AB, BC; for BK is equal to BC, four times the rectangle AB, BC, is quadruple of AK. But the gnomon STY was demonstrated to be four times of AK; therefore four times of that which is contained under AB, BC, is equal to the gnomon STY. To each of
[f] Cor. 4. 2. these add XH, which is equal[f] to the square of AC. Therefore four times the rectangle AB, BC, together with the square of AC, is equal to the gnomon STY, and the square XH. But the gnomon STY and the square XH make up the whole figure AEFD, which is the square of AD. Therefore four times the rectangle AB, BC, together with the square of AC, is equal to the square of AD; that is, of AB and BC added together in one line. Wherefore, if a right line, &c. Q. E. D.

Deduction.

Upon a given right line, as an hypothenuse, to describe a right-angled triangle, such that the hypothenuse, together with the less of the two remaining sides, shall be double of the greater of those sides.

The same by Algebra.

Let a equal the line AB, and let it be divided into any two parts e, f; then shall $4af + e^2 = \overline{a + f}^2$, whereof $4af$ is four times the rectangle AB, BC, e^2 the square of

AC, and $\overline{a+f}^2$ the square of AD. For $2af$ * $= a^2 +$ * 7. 2.
$f^2 - e^2$. Therefore $4af + e^2$ † $= a^2 + 2af + f^2 =$ † 4. 2.
$\overline{a+f}^2$, Q. E. D.

PROPOSITION IX.

THEOREM.

If a right line be any how cut into two equal and two unequal parts, then the squares of the two unequal parts are, together, double of the square of the half line, and the square of the intermediate part.

Let any right line AB be cut into two equal parts in the point C, and two unequal parts in the point D. The squares of AD, DB, together, are double to the squares of AC, CD.

For let[a] CE be drawn from the point C at right angles to AB, which make equal to AC or CB, and join EA, EB. Also through D let[b] DF be drawn parallel to CE, and FG through F parallel to AB, and draw AF. [a] 11. 1. [b] 31. 1.

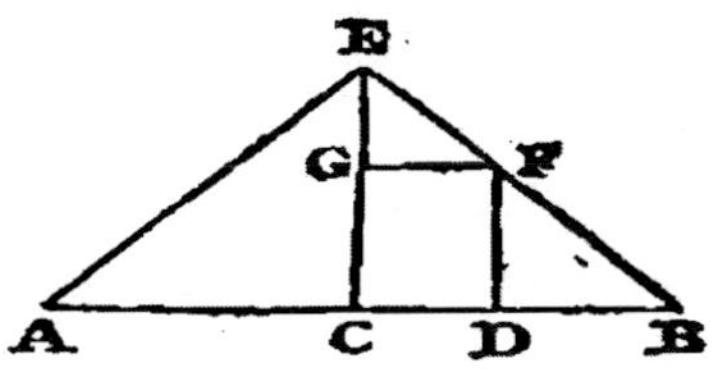

Then because AC is equal to CE, the angle EAC is equal[c] to the angle AEC; and because the angle at C is a right one, the other angles AEC, EAC, together, shall make[d] one right angle, and are equal to each other: each of them, therefore, is half a right angle. For the same reason are also CEB, CBE, each of them half right angles. Therefore the whole angle AEB is a right angle. And because the angle GEF is half a right angle, and EGF a right angle, for it is equal[e] to the interior and opposite angle ECB, the remaining angle EFG is half a right angle. Therefore the angle GEF is equal to the angle EFG, and also the side EG is equal[f] to the side GF. Again, because the angle at B is half a right one, and FDB is a right one, because it is equal to the inward and opposite angle ECB, the other angle BFD will be half a right angle; therefore the angle at B is equal to the angle BFD, and the side DF to the side DB. And since AC is equal to CE, the square of AC is equal to the square of CE; therefore the squares of AC, CE, together, are double to the square of AC; but the square of EA is equal[g] to the squares of AC, CE, together, since ACE is a right angle; therefore the [c] 5. 1. [d] 32. 1. [e] 29. 1. [f] 6. 1. [g] 47. 1.

square of EA is double of the square of AC. Again, since EG is equal to GF, therefore the squares of EG, GF, are double of the square of GF. But the square of EF is equal to the squares EG, GF, Therefore the square of EF is double the square of GF, but GF is
h 34. 1. equal[h] to CD, and so the square of EF is double to the square of DC. Wherefore the squares of AE and EF are double to the squares of AC and CD. But the square of AF is equal to the squares of AE and EF, because AEF is a right angle; therefore the square of AF is double of the squares of AC, CD. But the squares of AD, DF, are equal to the square of AF, because the angle ADF is a right angle; wherefore the squares of AD, DF, are double of the squares of AC, CD, and DF is equal to DB; wherefore the squares of AD, DB, are double of the squares of AC, CD. If therefore a right line, &c. Q. E. D.

The same by Algebra.

Put $a =$ AD, $b =$ DB, $c =$ AC, and $d =$ CD, then
* 4. 2. shall $a^2 + b^2 = 2c^2 + 2d^2$. For $a^2 + b^2$ * $= c^2 + d^2$
† 7. 2. $+ 2cd + b^2$. But $2cd + b^2$ † $= c^2 + d^2$, whence $a^2 + b^2 = 2c^2 + 2d^2$.‡ Q. E. D.

PROPOSITION X.

Theorem.

If a right line be cut into two equal parts, and produced to any point, the square of the whole line thus produced, and the part of it produced, shall be together double to the square of the half line, and of the line made up of the half and the part produced.

Let the right line be cut into two equal parts in the point C, and produced to the point D; the squares of AD, DB, together, are double to the squares of AC, CD, together.

a 11. 1. For draw[a] CE from the point C at right angles to AB,
b 31. 1. and make it equal to AC, or CB, and join AE, EB; likewise through E let EF be[b] drawn parallel to AD, and through D, DF parallel to CE. Then because the right

‡ This may be otherwise delivered and more easily demonstrated thus: the aggregate of the squares made of the sum and difference of two right lines is equal to the double of the squares made from those lines. For let the two right lines be a and b, then $a^2 + b^2 + 2ab = \overline{a+b}^2$, and $a^2 + b^2 - 2ab = \overline{a-b}^2$. These added together make $2a^2 + 2b^2$. Q. E. D.

line EF falls upon the parallels EC, FD, the angles CEF, EFD, are equal[c] to two right angles. Therefore the [c] 29. 1.
angles FEB, EFD, are less than two right angles. But right lines making with a third line angles together less than two right angles being infinitely produced will meet.[d] Wherefore EB, FD, produced, will meet to- [d] 12 Ax.
wards the parts B, D. Let them be produced and meet in the point G, and let AG be drawn.

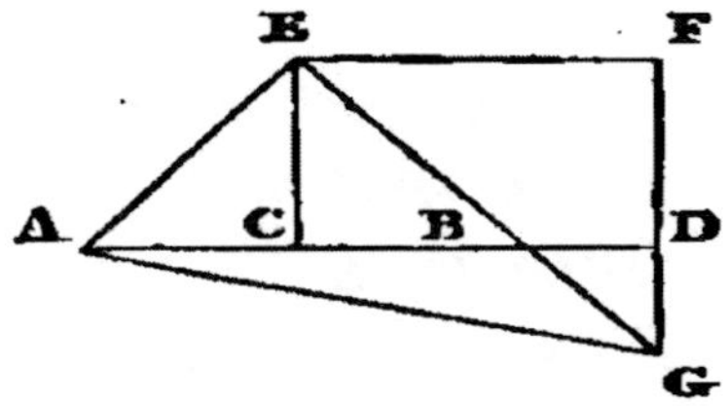

Then because AC is equal to CE, and the angle AEC shall be equal to the angle EAC.[e] But the angle at C is a [e] 5. 1.
right one; therefore the angle EAC or AEC is half a right one. By the same reasoning the angle CEB or EBC is half of a right one. Wherefore AEB is a right angle, and since EBC is half a right angle, DBG will[f] also be [f] 15. 1.
half a right angle, since it is vertical to CBE. But BDG is a right angle also, for it is equal to the alternate angle DCE. Wherefore the angle DBG is equal to the angle DGB. And thence in the triangle DBG the sides BD, DG, are equal. Again, because the right lines BD, EF, are parallel, and the right line EG falls upon them, the angle DBG will be equal to the angle GEF, and in the same manner GBD will be equal to EGD. Wherefore the angles GEF, EGF, are equal, and in the triangle FGE the side GF is equal[g] to the side EF. And [g] 6. 1.
since EC is equal to CA, and the square of EC equal to the square CA, therefore the squares of EC, CA, together, are double of the square CA. But the square of EA is equal[h] to the squares of EC, CA, and conse- [h] 47. 1.
quently double of the square of CA. Again, because EF is equal to GF, the square of GF also is equal to the square of FE. Wherefore the squares of GF, FE, are double to the square of FE. But the square of EG is[h] equal to the squares of GF, FE. Therefore the square of EG is double to the square of EF, but EF is equal to CD. Wherefore the square of EG shall be double to the square of CD. But it was proved that the square of EA is double of the square of AC; therefore the squares of AE, EG, are double of the squares of AC, CD. And the square of AG is equal[h] to the squares of AE, GE. But the squares of AD, GD, are equal[h] to the square of AG; therefore the squares of AD, DG, are double of the squares of AC, CD. But DG is equal to DB; therefore the squares of AD, DB, are double of the squares of AC, CD. Wherefore if a right line, &c. Q. E. D.

The same by Algebra.

Put a for the line AB, and b for the added line BD, then shall $b^2 + a^2 + 2ab + b^2 = \frac{1}{2}a^2 + \frac{1}{2}a^2 + 2ab +$
* Cor. 4. 2.
† 4. 2. $2b^2$. For $2 \cdot \frac{1}{2}a^2$* $= \frac{1}{2}a^2$, and $2 \cdot \overline{\frac{1}{2}a + b}^2$† $= \frac{1}{2}a^2 + 2b^2 + 2ab$. Q. E. D.

PROPOSITION XI.

Problem.

To cut a given right line into two parts so that the rectangle which is contained under the whole line and one part may be equal to the square made of the other part.

Let AB be a given right line. It is required to cut it so that the rectangle contained under the whole line and one part may be equal to the square made of the other part.

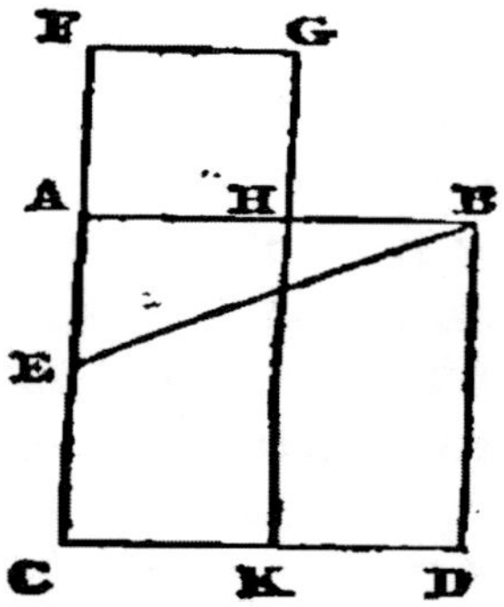

[a] 46. 1. Let the square ABCD be described[a] on AB, bisect AC in E, and draw BE; then CA being produced to F so that BE may be equal to EF; and on AF let the square AFGH be described, and GH produced to K. AB is cut in H so that the rectangle under AB, BH, is equal to the square of AH.

For since the right line AC is bisected in E, and AF is added directly thereto, the rectangle under CF, FA, together with the square
[b] 6. 2. of AE, shall be equal[b] to the square of EF; but EF is equal to EB; therefore the rectangle under CF, FA, together with the square of AE, is equal to the square
[c] 47. 1. made on EB. But the squares of AB, AE, are equal[c] to the square of EB, for the angle at A is a right one: therefore the rectangle under CF, FA, together with the square of AE, is equal to the squares of BA, AE; and if the square of AE, which is common, be taken away, the remaining rectangle under CF, FA, is equal to the square of AB. But FK is the rectangle under CF, FA; for AF is equal FG, and the square of AB is AD. Therefore the rectangle FK is equal to the square of AD. And if AH, which is common, be taken away, therefore the remaining FH is equal to the remaining HD. But HD is the rectangle under AB, BH, for AB is equal to BD, and FH is the square of AH. Therefore the rectangle AB, BH, shall be equal to the square of AH. And so the given right line AB is cut in H, so

that the rectangle under AB, BH, is equal to the square of AH. Which was to be done.

PROPOSITION XII.

THEOREM.

In obtuse angled triangles, the square of the side subtending the obtuse angle is greater than the squares, which are made by the sides containing the obtuse angle by twice the rectangle contained under one of the sides, which are about the obtuse angle, viz. that on which produced the perpendicular falls, and the line taken without between the perpendicular and obtuse angle.

Let ABC be an obtuse angled triangle, having the obtuse angle BAC, and draw from the point B the perpendicular BD to CA produced. The square of BC is greater than the squares of BA, AC, by twice the rectangle which is contained under CA, AD. For because the right line CD is any how cut in the point A, the square of CD shall be equal[a] to the squares of CA, AD, and to twice the rectangle AC, AD. To each of these equals add the square of DB. Therefore the squares of CD, DB, are equal to the squares of CA, AD, and twice the rectangle CA, AD. But the square of CB is equal[b] to the squares of CD, DB, for the angle at D is a right one, since BD is perpendicular, and the square of AB is equal[b] to the squares of AD, DB. Therefore the square of CB is equal to the squares of CA, AB, and twice the rectangle under CA, AD. Therefore the square of CB is greater than the squares of CA and AB by twice the rectangle contained under CA, AD. Therefore in obtuse angled triangles, &c. Q. E. D.

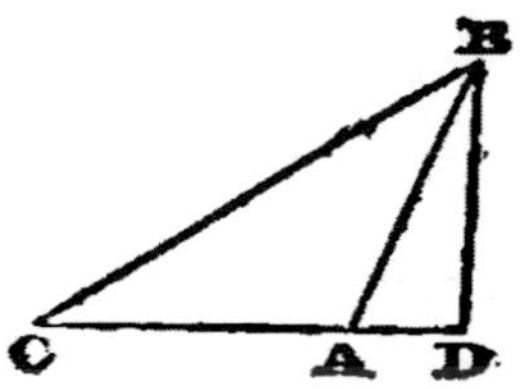

[a] 4. 2.

[b] 47. 1.

The same by Algebra.

Put a = CB, b = CA, c = AB, d = DB, and e = AD.

Then $a^2 = {}^*d^2 + \overline{b+e}^2$
$= {}^\dagger d^2 + b^2 + 2\,b\,e + e^2.$
$= b^2 + 2\,b\,e + c^2.$

* 47. 1.
† 4. 2.

PROPOSITION XIII.

Theorem.

In acute angled triangles, the square of the side subtending any of the acute angles is less than the squares of the sides containing the acute angle, by twice a rectangle under one of the sides about the acute angle, viz. on which the perpendicular falls, and the line assumed within the triangle from the perpendicular to the acute angle.

Let ABC be any acute angled triangle having an
[a] 12. 1. acute angle at B, and from the point A draw[a] AD perpendicular to BC. The square of AC is less than the squares of AB, BC, by twice the rectangle contained under BD, BC. For since the right line BC is any how cut in the point D, the
[b] 7. 2. squares of BD, BC, shall be equal[b] to twice a rectangle under CB, BD, together with the square of DC. And if the square of AD be added to both, then the squares of CB, BD, and DA, are equal to twice the rectangle under CB and BD, together with the squares of AD and DC. But the square of
[c] 47. 1. AB is equal to the squares of BD, DA,[c] for the angle at D is a right one. And the square of AC is equal[c] to the squares of AD, DC. Therefore the squares of CB and BA are equal to the square of AC together with twice the rectangle under CB and BD. Wherefore the square of AC only is less than the squares of CB and BA, by twice the rectangle under CB and BD. Therefore in acute angled triangles, &c. Q. E. D.

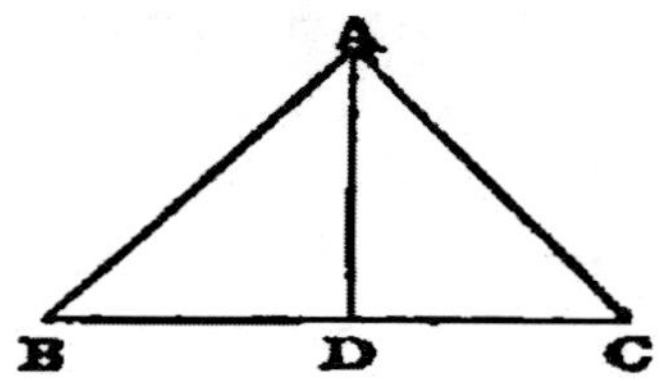

This proposition will hold true in obtuse and right angled triangles as well as acute, as may be perceived by the following demonstration.

Let ABC be an obtuse angled triangle, and the perpendicular fall without the triangle, as AD. For since BD is divided into two parts in the
[d] 7. 2. point C, the squares of BC, BD, are equal[d] to twice the rectangle of BC, BD, together with the square of DC. And if to each of these equals there be added the square of AD, the squares of CB, BD, and AD, will be equal to twice the rectangle of BC, BD, together with the sum of the squares of AD, DC. But the squares of BD, AD, are equal to the square

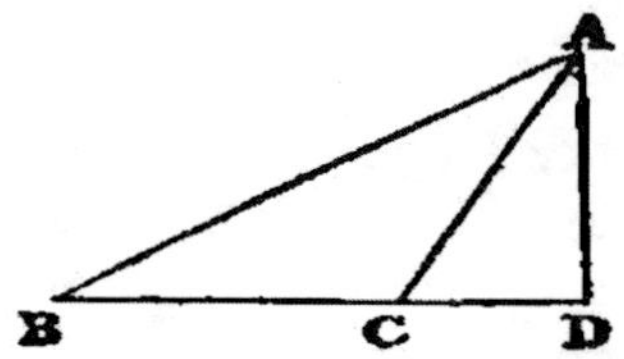

of AB, and the squares of AD, DC,[e] to the square of AC; whence the square of AC is less than the sum of the squares of BC, BA, by twice the rectangle BC, BD. [e] 47. 1.

Again, if the side AC be perpendicular to BC, then is BC the right line between the perpendicular and the acute angle at B; and it is manifest that the squares of AB, BC, are equal[f] to the square of AC and twice the square of BC. Q. E. D. [f] 47. 1.

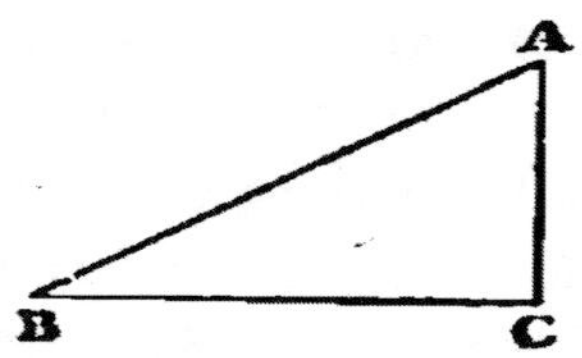

The same by Algebra.

Put a = BC, b = AB, c = AC, d = BD, e = DC, and f = AD.

Then $c^2 = e^2 + f^2$ *   * 47. 1.

but $c^2 + d^2 = 2cd + e^2$ †   † 7. 2.

$= 2cd + a^2 - f^2$

$\therefore a^2 + 2cd = c^2 + d^2 + f^2$

$= c^2 + b^2$ *

PROPOSITION XIV.

PROBLEM.

To describe a square equal to a given right lined figure.

Let A be the given right lined figure. It is required to describe a square equal thereto.

Describe[a] the right angled parallelogram BCDE equal to the right lined figure A. Then if BE is equal to ED, what was proposed will be done, for the square BD is described equal to the given right lined figure A. But if BE, ED, are unequal, produce one of them to F, and make EF equal to ED. Then BF being bisected in G, about which as a centre with the distance GB or GF, describe the semicircle BHF, and let DE be produced to H, and draw GH. Now because the right line BF is bisected at G, and divided into two unequal parts in E, the rectangle under BE, EF, together with the square of EG, will be equal[b] to the square of GF. But GF is equal to GH. Wherefore the rectangle under BE, EF, together with the square of GE, is equal to the square of GH. But the squares of HE, GE, are equal[c] to the square of GH: therefore the rectangle under BE, EF, together with the [a] 45. 1. [b] 5. 2. [c] 47. 1.

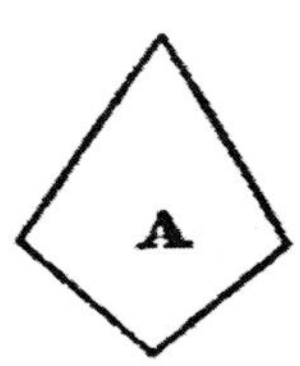

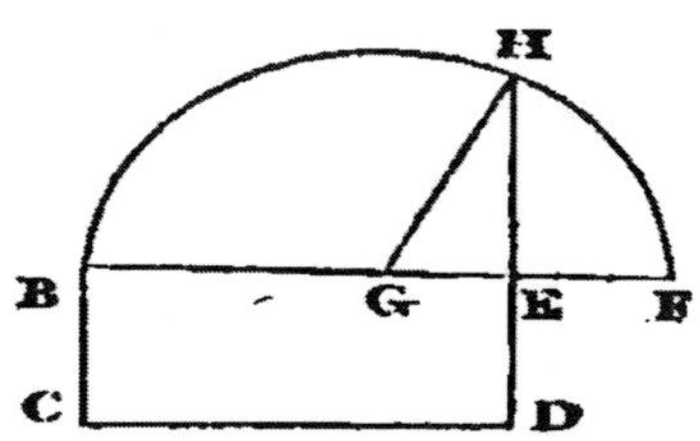

square of GE, is equal to the squares of HE, GE; take away the square of GE from both, then the remaining rectangle under BE, EF, is equal to the square of EH. But the rectangle under BE, EF, is the parallelogram BD, because EF is equal to ED: therefore the parallelogram BD is equal to the square of HE. But the parallelogram BD (by const.) is equal to the right lined figure A: therefore the right lined figure A will be equal to the square of EH. Which was to be done.

Deduction.

To find a line D, the square of which shall be equal to the sum of the rectangles AB, AC, BC: A, B, C, being three given lines.

In the demonstration of this, Dr. Keil, in his edition, has the words, "but if it be not, let either BE or ED be the greater, suppose BE, which let be produced to F," as if it was of any consequence, as Dr. Simson observes, whether the greater or less be produced; instead of these words there ought to be read, "But if they are not equal, produce one of them to F," as in the Oxford edition of Commandine.

EUCLID'S ELEMENTS.

BOOK III.

DEFINITIONS.

1. Equal circles are those of which the diameters are equal, or from the centres of which the right lines drawn to the circumferences are equal.

2. A right line is said to touch a circle, which, touching the circle, and produced, does not cut it.

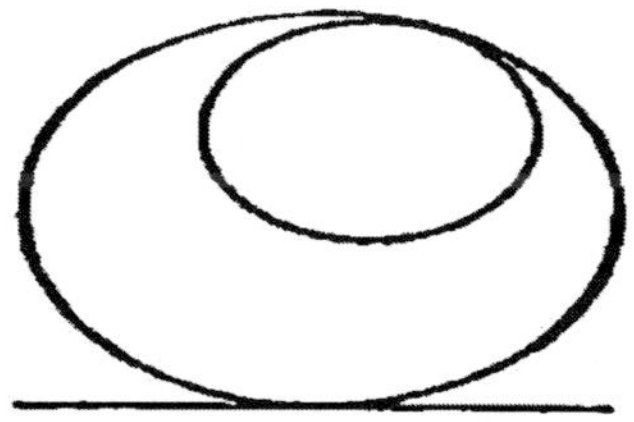

3. Circles are said to touch one another, which touching do not cut one another.

4. Right lines are said to be equally distant from the centre of a circle, when the right lines drawn from the centre perpendicular to them are equal.

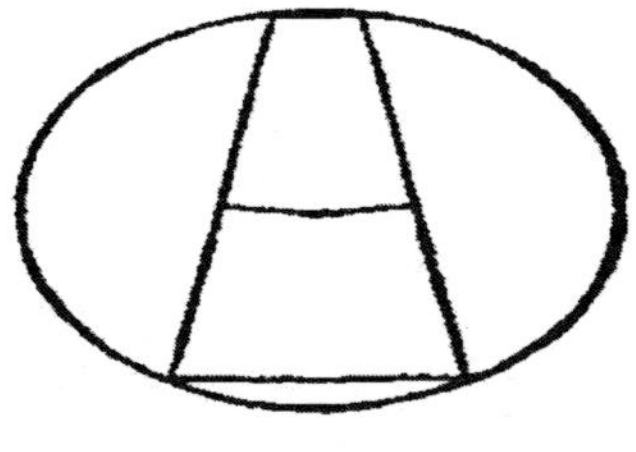

5. And that right line on which the greater perpendicular falls, is said to be further from the centre.

6. A segment of a circle is the figure contained both by the right line and the circumference of the circle it cuts off.

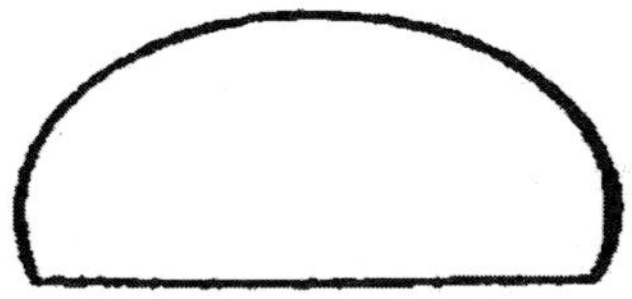

7. The angle of a segment is that which is contained by a right line and the circumference of the circle.

8. The angle in a segment is, when some point is taken in the circumference, and from it at the ends of a right line, which is the base of the segment, right lines are joined, the angle contained by the right lines being joined.

9. When the right lines containing an angle assume some circumference, the angle is said to stand upon the circumference.

10. The section of a circle is, when the angle is placed at the centre of a circle, the figure contained by the right lines containing the angle and the circumference between them.

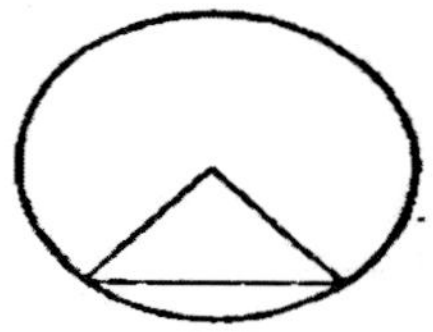

11. Similar segments of a circle are those in which the angles are equal to one another, or which contain equal angles.

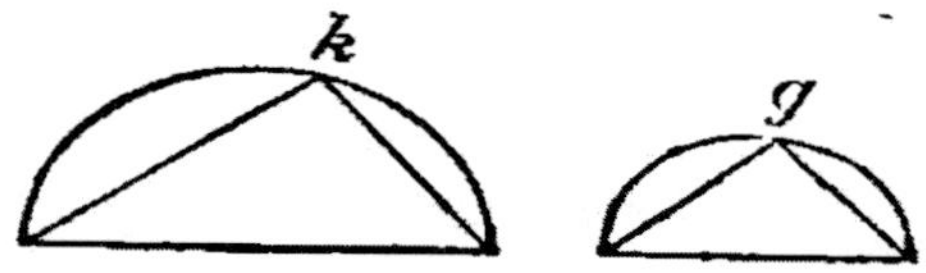

PROPOSITION I.*

PROBLEM.

To find the centre of a given circle.

Let ABC be the given circle; it is required to find the centre of the circle ABC. Draw in it anyhow the right line AB; which bisect[a] in the point D. From the point D draw[b] DC at right angles to AB, and produce it to E; and bisect[a] CE in F. Then is F the centre of the circle ABC. For if it be not; let G be the centre, if it be possible, and join GA, GD, GB. Therefore since DA is equal to DB, and DG common; the two AD, DG, are equal to the two DB, GD, each to each; and the base GA is equal to the base GB; for they are from the centre G. Therefore the angle ADG is equal[d] to the angle GDB. But when a right line standing on another right line makes the adjacent angles equal to one another, each of them is a right angle.[e] Therefore the angle GDB is a right angle; the angle FDB is also a right angle; therefore the angle FDB is equal to the angle GDB, the greater to the less, which is impossible. Wherefore G is not the centre of the circle ABC. In like manner it may be demonstrated that none other than F is the centre. Wherefore F is the centre of the circle ABC. Q. E. F.

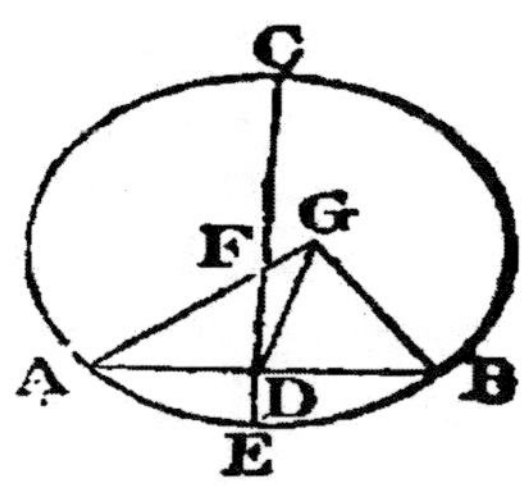

[a] 10. 1.
[b] 11. 1.
[d] 8. 1.
[e] 10. Def. 1.

COROLLARY.

If in a circle a right line bisects another right line at right angles, the centre of the circle shall be in the cutting line.

PROPOSITION II.

THEOREM.

If in the circumference of a circle, any two points are taken, the right line which joins them shall fall within the circle.

Let ABC be a circle; in its circumference take any two points A, B. The right line AB, which is drawn from A to B, falls within the circle. For in the right line AB

* Tacquet, and other authors, have proposed various methods for finding the centre of a circle; but none, I think, so simple in its operation as that of Euclid.

take any point E; join DA, DE, DB; and in DE produced, if necessary, take DF equal to DA or DB.
a 5. 1. Because DA is equal to DB, the angle DAB will be equal[a] to the angle DBA, and because the side AE of the trian-
b 16. 1. gle DAE is produced, the angle DEB will be greater[b] than the angle DAE; but the angle DAE is equal to the angle DBE; wherefore the angle DEB is greater than the angle DBE. But the greater side is subtended by the greater angle; wherefore also DF, which is taken equal to DB, is greater than DE. Therefore the point E necessarily lies between the points D, F. But because DB, DF, are equal, the point F will be at the circumference of the circle. Therefore the point E must fall within the circumference of the circle. In like manner it can be demonstrated that of every other point of the right line AB, between the points A, B, is within the circumference of the circle. If, therefore, in the circumference of a circle, &c. Q. E. D.

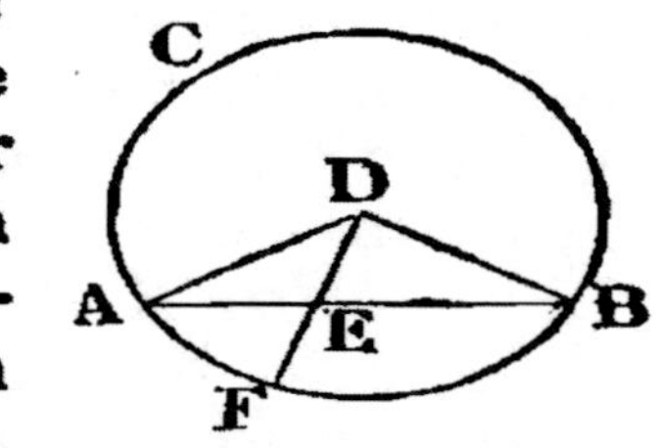

PROPOSITION III.

Theorem.*

If in a circle a right line drawn from the centre bisects another right line not drawn through the centre, it will also cut it at right angles; but if it cuts it at right angles, it shall also bisect it.

Let ABC be a circle, and in it a right line CD drawn through the centre, which bisects the right line AB, not drawn through the centre in the point F. It will cut it
a 1. 1. at right angles. Find the centre[a] of the circle ABC, which let be E, and join EA, EB. Therefore because AF is equal to FB, and FE common, the two AF, FE, are equal to the two BF, FE, and the base EA is equal
b 8. 1. to the base EB. Therefore also the angle[b] AFE will be equal to the angle BFE. But when a right line standing on a right line makes the adjacent angles

---

* The truth of this theorem is evident from a consideration of the first, for as the construction of that problem is effected by drawing a right line dividing the same into two equal parts, and from the point of bisection drawing another line perpendicular to the former; also, as it is clearly demonstrated that to assume any other point as the centre which is not in the perpendicular would be absurd, it follows conversely that the right line passing through the centre bisects another line not passing through the centre, *must* cut it at right angles, and, on the contrary, if it cut it at right angles, it *must* bisect it.

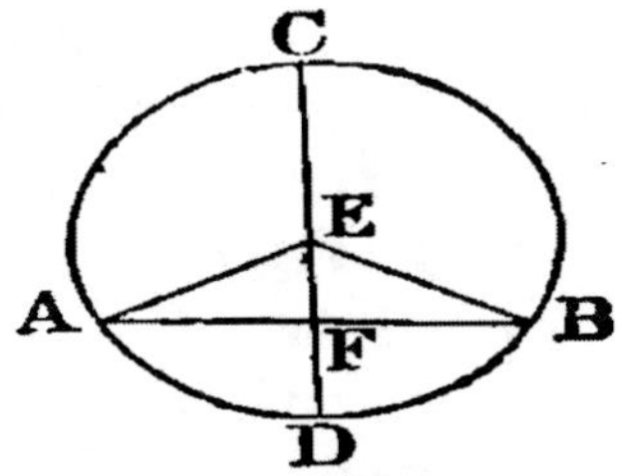

equal, then each of the equal angles is a right[c] angle; therefore AFE, BFE, are right angles. Wherefore the right line CD drawn through the centre, bisecting the right line AB not drawn through the centre, it will also cut it at right angles. But if CD cut AB at right angles, it also bisects it; that is, AF is equal to FB. For, by the same construction, because EA, which is from the centre, is equal to EB, the angle EAF will be equal to the angle[d] EBF, but the right angle AFE is also equal to the right angle BFE; therefore the two triangles EAF, EBF, have two angles equal, each to each, and one side equal to one side; namely, the side EF common to the two triangles, which is subtended by one of the equal angles. Therefore they will have the remaining sides equal to[e] the remaining sides, and AF will be equal to FB. If, therefore, in a circle a right line drawn through the centre bisect another right line which is not drawn through the centre, it will also cut it at right angles, and if it cut it at right angles, it will also bisect it. Q. E. D.

[c] Def. 10. 1.

[d] 5. 1.

[e] 26. 1.

Deduction.

If a right line drawn through the centre of a circle bisect any number of right lines which do not pass through the centre, the lines shall be parallel to one another.

PROPOSITION IV.

Theorem.

If in a circle two right lines cut another, which are not drawn through the centre, they shall not bisect one another.

Let ABC be a circle, and in it draw two right lines AC, BD, which cut one another in the point E, and are not drawn through the centre. They do not bisect each other. For if it be possible let them bisect each other, so that AE be equal to EC, and BE to ED, and find the centre[a] of the circle ABCD, which let be F, and join EF. Therefore because the right line FE drawn through the centre bisects another right line AC, which is not drawn through the centre, it will cut it at right[b] angles; wherefore FEA is a right angle. Again, because the right line FE

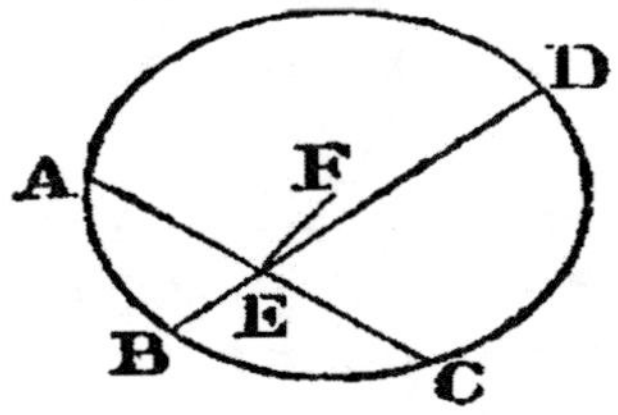

[a] 1. 1.

[b] 3. 1.

bisects another right line BD, which does not pass through the centre, it will also cut it at right angles; therefore FEB is a right angle. But FEA was shown to be a right angle. Wherefore the angle FEA will be equal to the angle FEB, the less to the greater, which is impossible. Therefore AC, BD, do not bisect each other. Wherefore if in a circle two right lines, &c. Q. E. D.

PROPOSITION V.

Theorem.

If two circles cut one another, they shall not have the same centre.

For let the two circles ABC, CDG, cut one another in the point C. They have not the same centre. For if it be possible, let E be the centre, and join EC, and in the circumference CGD take any point G, which is not common to both circumferences, EG joined will cut the circumference ACB in F. Because E is the centre of the circle ABC, EC will be equal to EF, and because E is the centre of the circle CDG, EC will be equal to EG; but EC was shown to be equal to EF: wherefore EF will be equal to EG, the less to the greater, which is impossible. Therefore the point E is not the centre of the circles ABC, CDG. Wherefore if two circles, &c. Q. E. D.

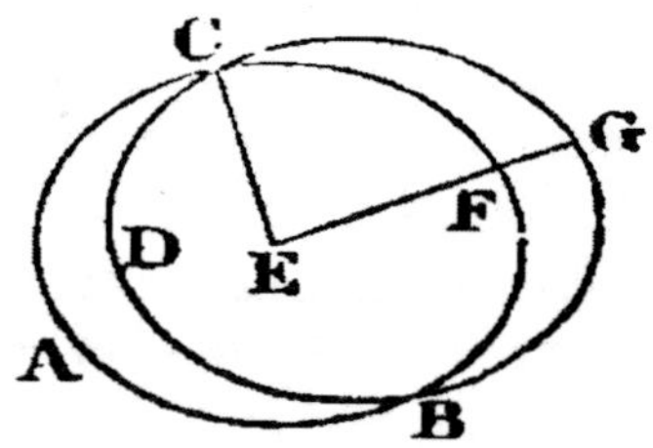

PROPOSITION VI.

Theorem.

If two circles touch one another internally, they shall not have the same centre.

For let the two circles ABC, CDE, touch one another internally in the point C. They have not the same centre. For if it be possible, let F be the centre, and join FC, and in the circumference ABC take any point B, which is not common to both circumferences. FB joined shall meet the circumference ECD in E. Therefore because F is the centre of the circle ABC, CF is equal to FB. Again, because F is the centre of the circle CDE, CF will be equal to FE. But CF was shown

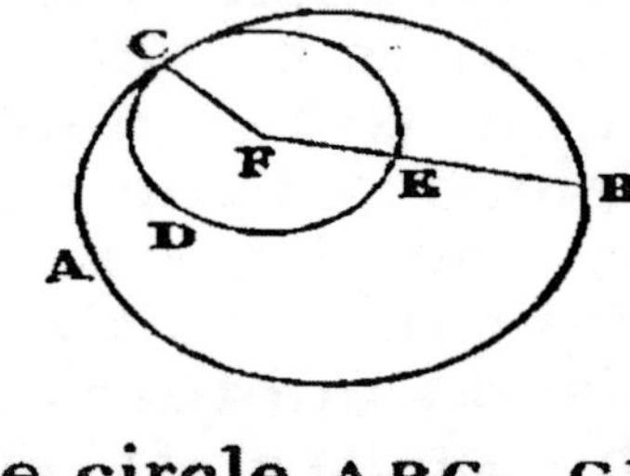

to be equal to FB; wherefore FE is also equal to FB, the less to the greater, which is impossible. Therefore the point F is not the centre of the circles ABC, CDE. Wherefore if two circles, &c. Q. E. D.

PROPOSITION VII.

THEOREM.

If in the diameter of a circle any point be taken which is not the centre of the circle, and from it any right lines be let fall in the circle, the greatest will be that in which the centre is, and the remainder the least; but of all the others, that which is nearer to that which passes through the centre is greater than that which is more remote, and only two equal right lines from the same point can be drawn in the circle one on each side of the least.

Let ABCD be a circle whose diameter is AD, and in AD take any point F, which is not the centre of the circle; and let E be the centre of the circle, and from the point F in the circle ABCD draw any right lines FB, FC, FG: FA is the greatest, and FD the least; but of the others, FB is greater than FC, and FC greater than FG. For join BE, CE, GE; and because two sides of every triangle are greater than the third; BE, EF, will be greater than BF. But AE is equal to BE; wherefore BE, EF, are equal to AF; AF is therefore greater than FB. Again, because BE is equal to CE, and FE common, the two BE, EF, are equal to the two CE, EF, but the angle BEF is greater than the angle CEF; therefore the base [a]BF is greater than the base FC. For the same reason CF is also greater than FG. Again, because GF, FE, are greater[b] than GE, but GE is equal to ED; GF, FE, will be greater than ED, take away the common part FE; therefore the remainder GF is greater than the remainder FD. Therefore FA is the greatest, and FD the least; also BF is greater than FC, and FC than FG. And from the point F only two equal right lines can be let fall into the circle ABCD, one on each side of the least FD. For at the line EF at the given point E in it, make[c] the angle HEF equal to the angle FEG; and join FH. Therefore because GE is equal to EH, and EF common, the two GE, EF, are equal to the two HE, EF, and the angle GEF is equal

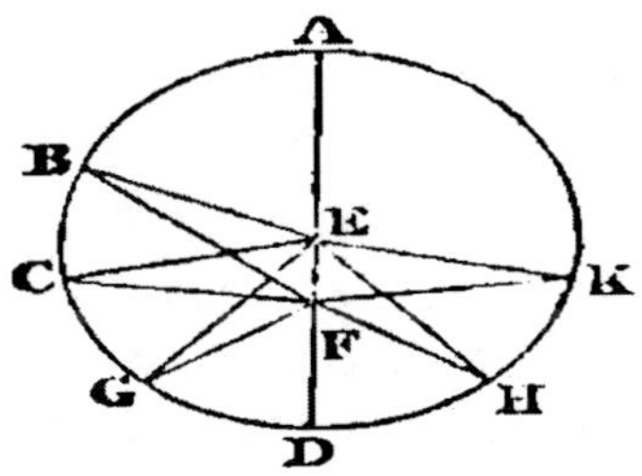

[a] 24. 1.
[b] 20. 1.
[c] 23.

[d] 4. 1. to HEF; therefore the base FG[d] will be equal to the base FH. From the point F no other right line can be let fall in the circle equal to FG. For if it be possible let FK fall, and because FK is equal to FG, and FH is equal to FG, FK will be also equal to FH, viz. a right line nearer to that which passes through the centre is equal to that which is more remote, which is impossible. If, therefore, in the diameter of a circle, &c. Q. E. D.

Deduction.

If two equal right lines in a circle meet in a point which is not the centre, then the right line which passes through the centre and that point bisects the angle contained by the two equal right lines.

PROPOSITION VIII.

Theorem.

If without a circle any point be taken, and from it any right lines be drawn to the circle, one of which passes through the centre, and the others any how; of those which fall on the concave circumference the greatest is that which passes through the centre; and of the others, that which is nearer to that which passes through the centre is greater than that which is more remote; and of those which fall on the convex circumference, that is the least which lies between the point and the diameter; and of the others, that which is nearer to the least is less than that which is more remote, and from that point only two equal right lines can be drawn to the circle, one on each side of the least line.

Let ABC be a circle, and without it take any point D; and from it draw to the circle certain right lines DA, DE, DF, DC; and let DA pass through the centre. Of those which fall on the concave circumference, AEFC, DA is the greatest, which passes through the centre, and the least is that which lies between the point D and the diameter AG; namely, DG; but DE is greater than DF, DF greater than DC, and of those which fall on the convex circumference HLKG, that which is nearer to DG the least is always less than that which is more remote; that is, DK is less than DL, and DL than DH. For find the centre[a] of the circle ABC which let

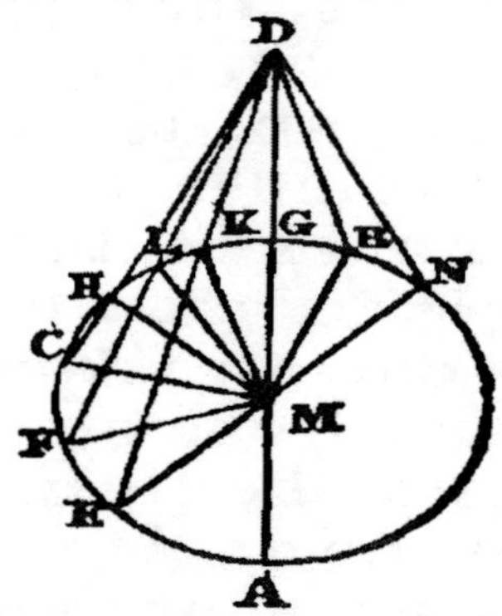

[a] 1. 3.

be M, and join ME, MF, MC, MH, ML, MK. And because AM is equal to ME, and MD is common; wherefore AD is equal to EM, MD; but EM, MD, are greater[b] than ED. [b] 20. 1.
Wherefore AD is also greater than ED. Again, because ME is equal to MF, add MD, which is common, EM, MD, will be equal to MF, MD, but the angle EMD is greater than the angle FMD; therefore the base ED will be greater than the base FD. In like manner we may demonstrate that FD[c] is also greater than CD. There- [c] 24. 1.
fore DA is the greatest, but DE is greater than DF, and DF than DC. Moreover, because MK, KD, are greater than MD, and MK is equal to MG, the remainder KD[d] [d] Ax. 4. 1.
will be greater than the remainder GD: wherefore GD is less than KD. And because in one side MD of the triangle MLD two right lines are drawn within it, viz. MK,[e] KD, these will be less than ML, LD, of which MK [e] 21. 1.
is equal to ML; the remainder, therefore, DK, is less than the remainder DL. In like manner we may show that DL is less than DH. Wherefore DG is the least, but DK less than DL, and DL less than DH. Also only two equal right lines can be drawn from the point D on each side of the least line. Make at the right line MD at the given point M in it, the angle DMB equal to the angle KMD,[f] and join DB. Therefore because MK is equal to [f] 23. 1.
MB, and MD common, the two KM, MD, are equal to the two BM, MD, each to each, and the angle KMD is equal to the angle BMD; the base, therefore, DK,[g] is equal to [g] 4. 1.
the base DB. From the point D no other right line can fall on the circumference equal to DK. For if it be possible, let DN fall, and because DK is equal to DN, and DK to DB, DB will be also equal to DN; that is, that which is nearer is equal to that which is more remote, which has been proved to be impossible. Therefore if without a circle, &c. Q. E. D.

PROPOSITION IX.*

THEOREM.

If within a circle any point be taken, and from it more than two equal right lines are drawn to the circumference, the point so taken will be the centre of the circle.

For within the circle ABC, take any point D, and from the point D let more than two equal right lines DA, DB,

* An affirmative demonstration may be, and is, given to this in many editions of the Elements; the present one, although it possesses not that advantage, I deemed preferable both for conciseness and simplicity.

DC, be drawn to the circle. The point D which is taken is the centre of the circle ABC. For if not, if it be possible, let E be the centre, and join DE in F, and produce it to G; wherefore FG is the diameter of the circle ABC. Therefore, because in FG, the diameter of the circle ABC, any point D is taken which is not

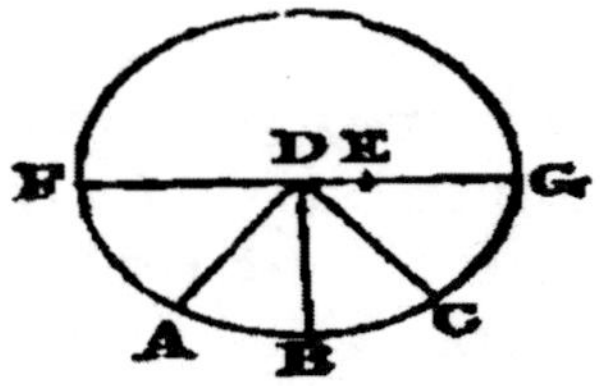

the centre of the circle ABC, DG will be the greatest,
a 7. 3. and DC[a] greater than DB, and DB than DA; but DB,[b]
b Ex hyp. DC, DA, are also equal, which is impossible; wherefore E is not the centre of the circle ABC. In like manner we can show that no other point than D is the centre. Wherefore D will be the centre of the circle BC. Q. E. D.

PROPOSITION X.

Theorem.

One circle cannot cut another in more points than two.

For if it be possible, let the two circles ABC, DBC, cut one another in the B, C, E. Join CB, CE. And let the right line CB be bisected in the point G, and from the point G draw the perpendicular right line GH. Let CE also be bisected in the point K, and from the point K draw the perpendicular KL. The two right lines GH, KL, drawn perpendicular to the two

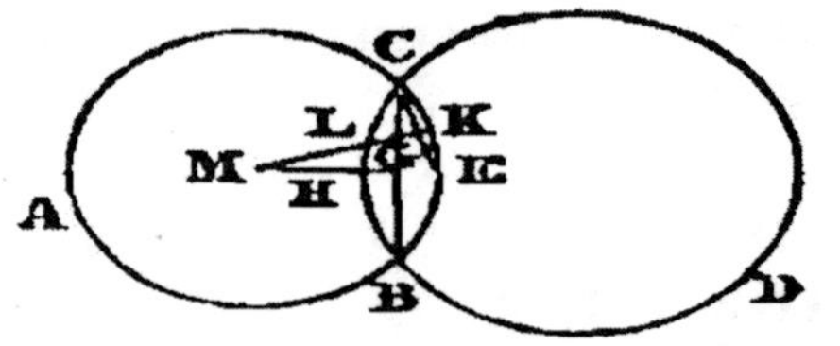

CB, CE, which are not parallel, are themselves not parallel to one another. Therefore they will meet. Let them meet at M. Now since the points, B, C, are at the cir-
a 2. 3. cumference of the circle ABC, the right line CB[a] will be within the circle. But the right line HG bisects the right line CB described in the circle ABC, and
b Per cons. it is at right[b] angles to it. Therefore the centre of
c Cor. 1. 3. the circle ABC will be in the right line GH.[c] For the same reason the centre of the circle ABC will be in the right line KL, which bisects the line CE drawn in a circle and at right angles. Therefore the centre of the circle ABC is a point common to the two right lines GH, KL. But of these right lines, M is the only common point of meeting. The point M is, therefore, the centre of the circle ABC. But in like manner we show that the same point M is the centre of the circle DBC, at whose circumference there are three points
Ex hyp. B, C, E,[d] in common with the circumference of the other

ABC. Therefore M is the common centre of the two circles ABC, DBC, which cut one another; which is absurd.[e] Therefore the two circles ABC, DBC, do not cut one another at the points B, C, E. For the same reason, neither can they do it in any other three. Therefore one circle, &c. Q. E. D. [e] 5. 3.

PROPOSITION XI.

THEOREM.

If two circles touch one another inwardly, and their centres be taken, the right line which joins their centres being produced, will fall at the point of contact of the circles.

For let the two circles ABC, ADE, touch one another internally at A, and find the centre of the circle ABC, which let be F, and G the centre of the circle ADE. The right line drawn from the point F to G, if produced, will meet at the point of contact A. For if not, if it be possible, let it fall as FGDH, and join AF, AG. Therefore, because AG,[a] GF, are greater than AF, that is, than FH, for FA is equal to FH, both being from the same centre. Take away the common part FG, therefore the remainder AG is greater than the remainder GH, but AG is equal to GD; wherefore GD is greater than GH, the less to the greater, which is impossible. Therefore the right line drawn from the point F to G does not fall beyond the point of contact A, wherefore it must necessarily fall in it. Wherefore if two circles, &c. Q. E. D. [a] 20. 1.

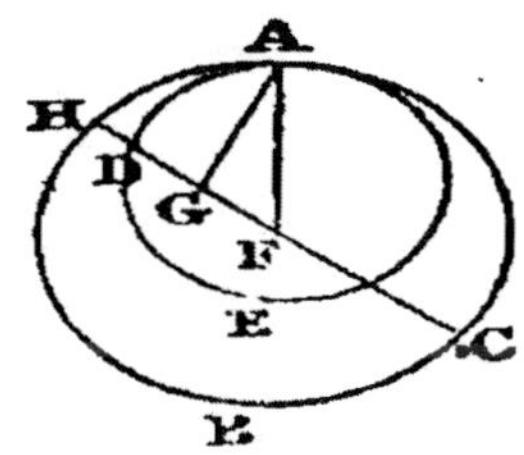

PROPOSITION XII.

THEOREM.

If two circles touch one another externally, the right line joining their centres will pass through the point of contact.

For let the circles ABC, ADE, touch one another externally at the point A, and find the centre of the circle ABC, which let be F, and G the centre of the circle ADE. The right line drawn from the point F to G will pass through the point of contact A. For if it does not, if it be possible, let it fall as FCDG,

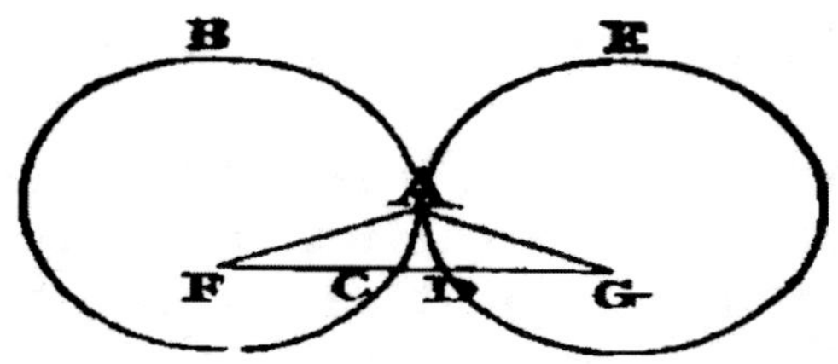

and join FA, AG. Because, therefore, F is the centre of the circle ABC, AF will be equal to FC. Again, because G is the centre of the circle ADE, AG will be equal to GD. But AF was shown to be equal to FC; FA, AG, are therefore equal to FC, DG; therefore the whole FG is greater
[a] 20. 1. than FA, AG, but it is also[a] less, which is impossible. Therefore the right line drawn from the point F to G does not pass elsewhere than through the point of contact A; wherefore it must necessarily pass through it. If therefore two circles, &c. Q. E. D.

PROPOSITION XIII.

Theorem.

One circle cannot touch another in more points than one, whether it touches it on the inside or outside.

For if it be possible let the circle ABDC touch the circle EBFD in the first place in the inside in more points than one B, D. And find G the centre of the circle ABDC, and H the centre of the circle EBFD. The right line drawn from G to H

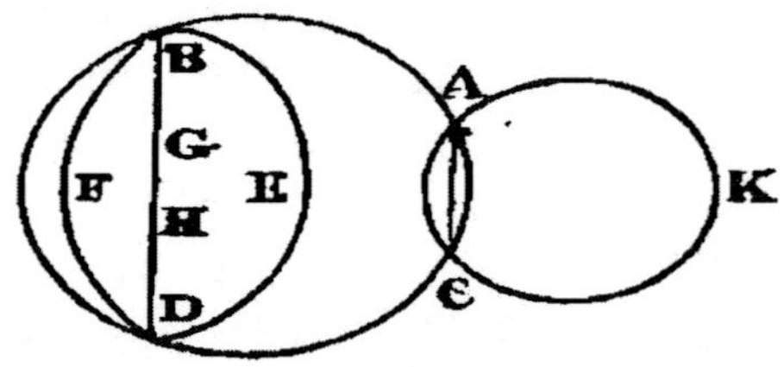

[a] 11. 3. will pass through the points B,[a] D. Let it fall as BGHD. And because G is the centre of the circle ABDC, BG is equal to GD; therefore BG is greater than HD; much more therefore is BH greater than HD. Again, because H is the centre of the circle EBFD, BH is equal to HD. But it has been shown to be much greater than it, which is impossible. One circle, therefore, cannot touch another internally in more points than one. Neither can it externally. For if it be possible, let the circle ACK touch the circle ABDC externally in more points than one in A, C, and join AC. Therefore because any two points are taken in each of the circumferences of the circles ABDC, ACK, the right
[b] 2. 3. line joining these points[b] shall pass within each of the circles. But because it falls within ABDC, it must fall
[c] 3 Def. 3. without ACK,[c] which is absurd. Therefore one circle, &c. Q. E. D.

PROPOSITION XIV.

THEOREM.*

In a circle equal right lines are equally distant from the centre, and those which are equally distant from the centre are equal to another.

Let ABDC be a circle, and in it the two equal right lines AB, CD. They are equally distant from the centre. For find the centre of the circle ABDC, which let be E, and from it draw EF, EG, perpendicular to AB, CD, and join AE, EC. Therefore because the right line EF drawn through the centre cuts the right line AB, which is not drawn through the centre at right angles, it shall bisect[a] it; wherefore AF is equal to FB, and consequently AB is double of AF. For the same reason CD is also double of CG, and AB is equal to CD; therefore also AF is equal to CG, and because AE is equal to EC, the square of AE will be equal to the square of EC, but the squares[b] of AF, FE, are equal to the square of AE; for the angle at F is a right angle, but the squares of EG, GC, are equal to the square of EC, for the angle at G is a right angle. Therefore the squares of AF, FE, are equal to the squares of CG, GE, of which the square of AF is equal to the square of CG, for AF is equal to CG. Therefore the remaining square described on FE is equal to the remaining square described on EG, and consequently FE is equal to EG. But in a circle right lines are said to be equally distant from the centre when the perpendiculars drawn to them from the centre are equal. Wherefore AB, CD, are equally distant from the centre. But if AB, CD, be equally distant from the centre; that is, if FE be equal to EG, AB is equal to CD. For the same construction being made, it may be shown in like manner that AB is double of AF; and CD the double of CG. And because AE is equal to EC, the square of AE will be equal to the square of EC, and the squares of EF, FA, are equal to the square of AE. Therefore the squares EF, FA, are equal to the squares EG, GC, of which the square of EF is

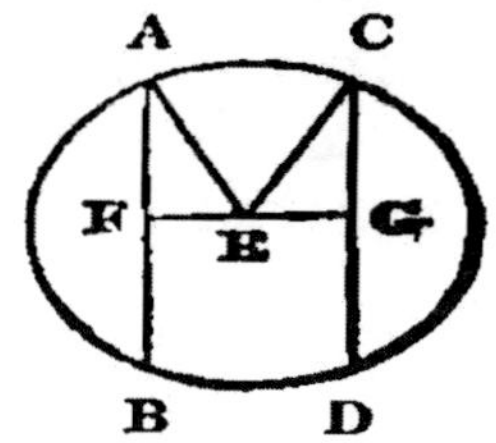

[a] 3. 3.

[b] 47. 1.

* Legendre's demonstration of this, as given in his Elements of Geometry, is deficient; for he says, bisect the chords AB, CD, by perpendiculars EF, EG; but he has no where proved that EF, EG, bisecting the lines AB, CD, *will* be perpendiculars; with this exception, his method is much shorter and preferable to that of Euclid.

equal to the square of EG, for EG is equal to EF; therefore the remaining square of AF is equal to the remaining square of CG; wherefore AF is equal to CG. But AB is the double of AF, and CD the double of CG. Therefore in a circle, &c. Q. E. D.

Deductions.

1. If two isosceles triangles be of equal altitudes, and have a side of the one equal to a side of the other, then shall their bases be equal.

2. There can only be drawn two equal right lines in a circle which are parallel to one another.

PROPOSITION XV.

Theorem.*

In a circle the greatest line is the diameter, but of the other lines, that which is nearer to that which passes through the centre is greater than that which is more remote.

Ler ABCD be a circle, of which AD is the diameter, and E the centre, and let BC be nearer to the diameter AD, but FG more remote, AD is the greatest, and BC greater than FG. For draw EH, EK, from the centre E, perpendicular to BC, FG. And because BC is nearer to that which passes through the centre, and FG more remote, EK will be greater than EH.[a] Make EL equal to EH, and through L draw LM, at right angles to EK, and produce it to N. And join EM, EN, EF, EG.

a 4. 3.

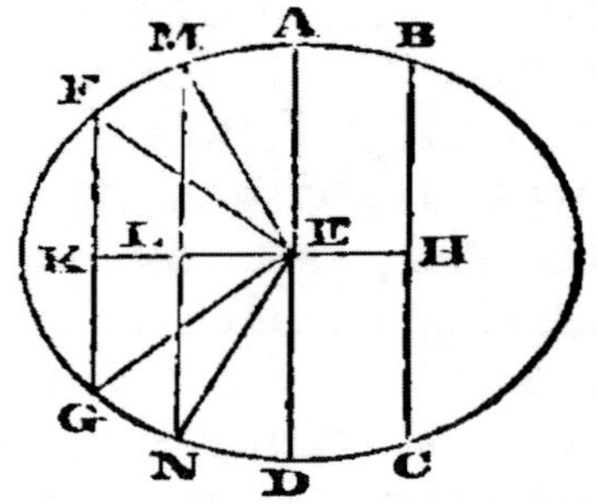

Therefore because EH is equal to EL, MN will be also equal to BC.[a] Again, because AE is equal to EM, and DE to EN, AD will also be equal to ME, EN; but ME, EN, are greater than MN. Wherefore also AD is greater than MN, and MN is equal to BC. Wherefore AD will be greater than BC. Because the two EM, EN, are equal to the two FE, EG, and the angle MEN greater than the angle FEG, the base MN will be greater than the base FG. But MN has been shown to be equal to

* The converse of this proposition is not added, as it is never used in any part of the Elements: it was necessary to prove in the 14th that right lines equally distant from the centre are equal to one another, because it is employed in this proposition.

BC; wherefore also BC is greater than FG. Therefore AD, the diameter, is the greatest, and BC is greater than FG. Wherefore in a circle, &c. Q. E. D.

PROPOSITION XVI.

THEOREM.

(Extracted from Dr. Simson's Edition.)

The right line drawn at right angles to the diameter of a circle from the extremity of it falls without the circle; and no right line can be drawn between that right line and the circumference from the extremity so as not to cut the circle; or, which is the same thing, no right line can make so great an acute angle with the diameter at its extremity, or so small an angle with the right line at right angles to it, as not to cut the circle.

Let ABC be a circle, the centre of which is D, and the diameter AB, the straight line drawn at right angles to AB from its extremity A, shall fall without the circle. For if it does not, let it fall if possible within the circle, as AC, and draw DC to the point C, where it meets the circumference. And because DA is equal to DC, the angle DAC is equal to the angle ACD;[a] but DAC is a right angle, therefore ACD is a right angle, and the angles DAC, ACD, are therefore equal to two right angles, which is impossible.[b] Therefore the straight line drawn from A at right angles to BA does not fall within the circle. In the same manner it may be demonstrated, that it does not fall upon the circumference, therefore it must fall without the circle, as AE. And between the same straight line AE, and the circumference, no straight line can be drawn from the point A which does not cut the circle. For, if possible, let FA be between them, and from the point D draw DG[c] perpendicular to FA, and let it meet the circumference in H. And because AGD is a right angle, and DAG less than a right angle, DA is greater than DG,[d] but DA is equal to DH. Therefore DH is greater than DG, the less than the greater, which is impossible. Therefore no straight line can be drawn from the point A between AE and the circumference which does not cut the circle, or which amounts to the

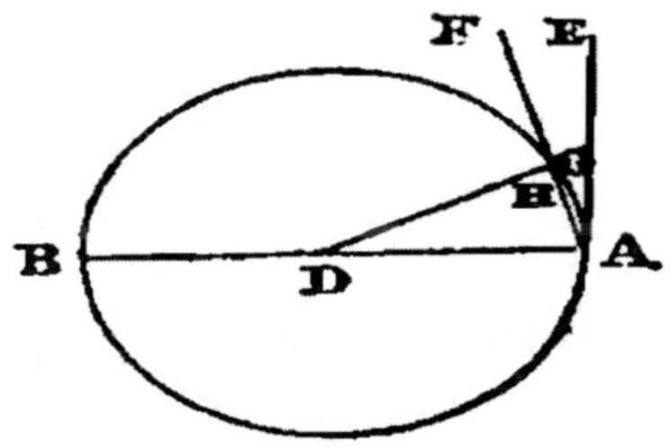

[a] 5. 1.

[b] 17. 1.

[c] 12. 1.

[d] 19. 1.

same thing, however great an acute angle a straight line makes with the diameter at the point A, or however small an angle it makes with AE, the circumference passes between that straight line and the perpendicular AE. And this is all that is to be understood when in the Greek text, and translations from it, the angle of the semi-circle is said to be greater than any acute rectilineal angle, and the remaining angle less than any rectilineal angle. Q. E. D.

Deductions.

1. The right line which is drawn at right angles to the diameter of a circle from the extremity of it touches the circle, and that it touches only in one point.

2. To describe a circle, which shall touch a given right line in a given point and also touch a given circle.

PROPOSITION XVII.*

Problem.

From a given point to draw a right line which shall touch a given circle.

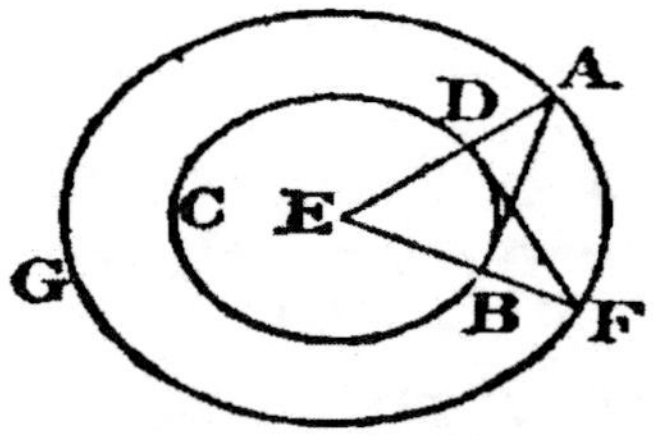

Let A be a given point, and BCD a given circle, it is required from the given point A to draw a right line which
[a] 1. 3. shall touch the circle BCD. Find E,[a] the centre of the circle, and join AE, which shall meet the circumference BDC in D; from the centre E, with the distance EA, describe the circle AFG, and from the point
[b] 11. 1. D draw DF,[b] at right angles to EA, which shall meet the circumference AFG in F; also join EF, which shall meet the circumference CDB in B; lastly join AB. From the point A, AB is drawn, which touches the circle BCD. For because E is the centre of the circles BCD, AFG, EA shall be equal to EF, and ED to EB. Therefore the two AE, EB, are equal to the two FE, ED, and they contain a common angle, namely, the angle at E. Wherefore the base DF is equal to the base AB, and the triangle DEF to the triangle EBA; also the remaining

* A better practical solution of this problem may be effected by means of the thirty-first proposition of this Book.

angles of the one to the remaining angles of the other.[c] [c 4. 1.]
Therefore the angle EBA is equal to the angle EDF, but EDF is a right angle. Wherefore EBA is a right angle, and EB is drawn from the centre. But the right line which is drawn at right angles from the extremity of the diameter of a circle touches the circle. Wherefore AB touches the circle. And from the given point A a right line AB is drawn, which touches the circle BCD. Q. E. F.

Deduction.

To draw a right line which shall be a tangent to two given circles not being one within the other.

PROPOSITION XVIII.

THEOREM.

If a right line touches a circle, and from the centre a right line be drawn to the point of contact, it shall be perpendicular to the touching line.

For let any right line DE touch the circle ABC, in the point C, and find F, the centre of the circle ABC, from which draw FC to C; FC is perpendicular to DE. For if it be not, from the point F, draw FG perpendicular to DE. Therefore because FGC is a right angle, GCF will be an acute angle,[a] [a 17. 1.]
and consequently the angle FGC is greater than the angle FCG. But the greater side subtends the greater angle.[b] [b 19. 1.]

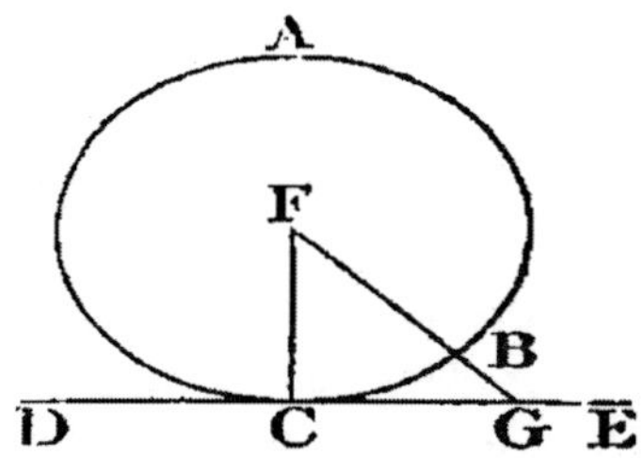

Wherefore FC is greater than FG. But FC is equal to FB: wherefore FB is greater than FG, the less than the greater, which is impossible. Therefore FG is not perpendicular to DE; in like manner we show that no other is so besides FC. Wherefore FC is perpendicular to DE. If therefore a right line, &c. Q. E. D.

Deductions.

1. Two right lines which touch the circumference of a circle in the opposite extremities of the diameter are parallel to one another.

2. If two parallel right lines touch a circle, they must touch it in the opposite extremities of the diameter, neither can more than two parallel right lines touch the same circle.

PROPOSITION XIX.

Theorem.

If a right line touches a circle, and from the point of contact a right line be drawn at right angles to the touching line, the centre of the circle shall be in that line.

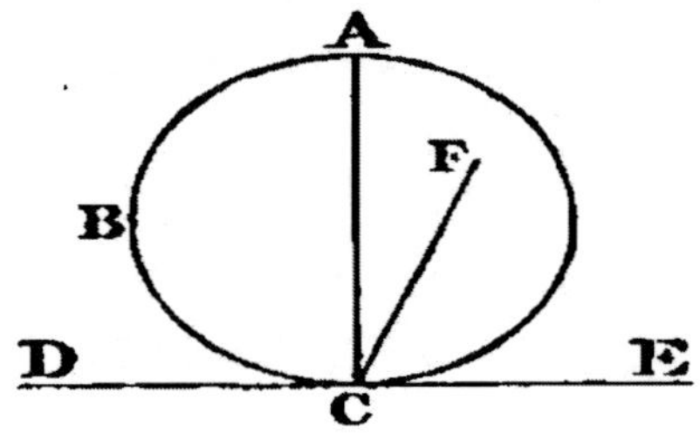

For let any right line DE touch the circle ABC in C, and from the point C draw CA at right angles to DE. The centre of the circle is in AC. For if not, if it be possible, let F be the centre, and join CF. Therefore because the right line DE touches the circle ABC, and from the centre FC is drawn to the
a 18. 3. point of contact, FC will be perpendicular to DC.[a] Therefore FCE is a right angle, but ACE is also a right angle; wherefore the angle FCE is equal to the angle ACE, the less to the greater, which is impossible. Therefore F is not the centre of the circle ABC. In like manner we show that it is not in any other than AC. Wherefore if a right line, &c. Q. E. D.

PROPOSITION XX.

Theorem.

In a circle the angle at the centre is double of that at the circumference, when they have the same circumference for their base.

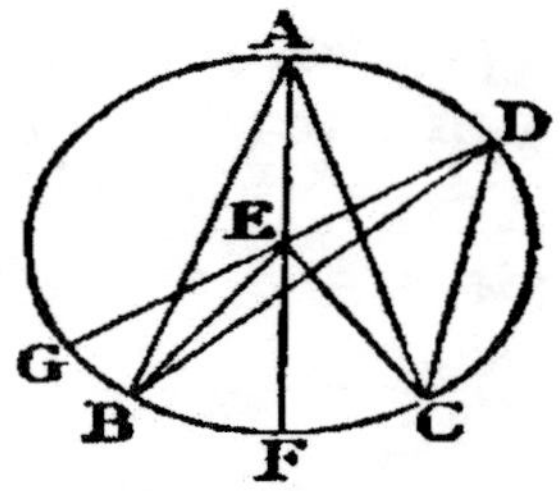

Let ABC be a circle, and BEC an angle at the centre, and BAC an angle at the circumference, which have the same circumference BC for their base. The angle BEC is double of the angle BAC. For join AE, and produce it to F. Therefore because EA is equal to EB, the angle
a 5. 1. EAB is equal to the angle EBA.[a] Therefore the angles EAB, EBA, are double of the angle EAB, but the angle BEF is equal
b 32. 1. to the angles EAB, EBA:[b] wherefore the angle BEF is double of the angle EAB. For the same reason the angle FEC is double of the angle EAC. Therefore the whole BEC will be double of the whole BAC. Again, let the centre E be without the angle BDC, join DE, and produce it to G. In like manner, we show that the angle GEC is double of the angle GDC,

of which GEB is double of GDB. Wherefore the remainder BEC is double of the remainder BDC. Therefore in a circle the angle, &c. Q. E. D.

Deduction.

If two chords of a given circle intersect each other, the angle of their inclination is equal to half the angle at the centre, which stands on an arc equal to the sum or difference of the arcs intercepted between them, according as they meet within or without the circle.

PROPOSITION XXI.

THEOREM.

In a circle the angles which are in the same segment are equal to one another.

Let ABCDE be a circle, and the angles BAD, BED, in the same segment, BACD. These angles are equal to one another. For find the centre of the circle ABCDE, which let be F, and join BF, FD. Because the angle BFD is at the centre, and the angle BAD at the circumference, also they have the same circumference BCD for

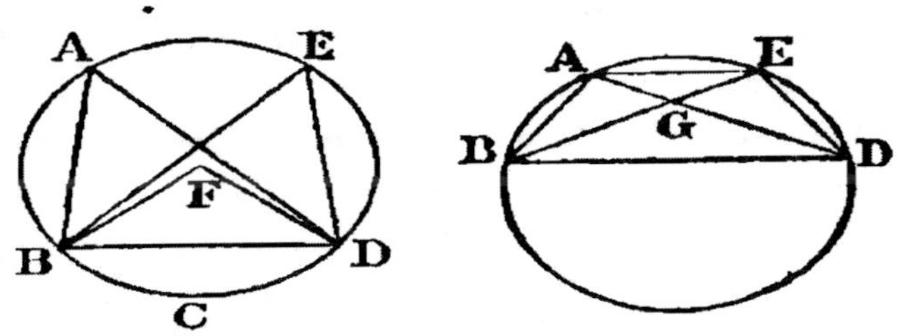

their base, the angle BFD[a] will be double of the angle BAD. For the same reason, the angle BFD is double of the angle BED. Wherefore the angle BAD will be equal to the angle BED. If the angles BAD, BED, are in a segment less than a semicircle, draw AE, and all the angles of the triangle ABG will be equal to all the angles of the triangle DEG,[b] and the angles ABE, ADE, are equal, as have been demonstrated, and the angles AGB,[c] DGE, are also equal, for they are vertically opposite. Wherefore also the remaining angle BAG will be equal to the remaining angle GED. Therefore in a circle, &c. Q. E. D.

[a] 20. 3.
[b] 32. 1.
[c] 15. 1.

Deductions.

1. If from a given point within a circle, which is not the centre, right lines be drawn to the circumference, making with each other equal angles, the two, which are nearest to the diameter passing through the

given point, shall cut off a greater circumference than the two, which are more remote.

2. Given the segments of the base of a triangle made by a perpendicular drawn from the vertex and the vertical angle to construct the triangle.

PROPOSITION XXII.

Theorem.

The opposite angles of quadrilateral figures which are inscribed in a circle are equal to two right angles.

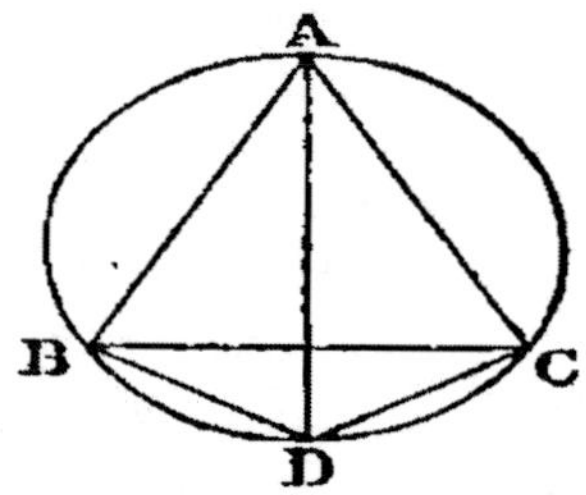

Let ABDC be a circle, and ABDC a quadrilateral figure in it. Then any two opposite angles of
a 32. 1. it are equal[a] to two right angles; join AD, BC. Therefore because the three angles of every triangle are equal to two right angles, the angles CAB, ABC, CBA, are equal to two right angles. But the
b 21. 3. angle ABC is equal[b] to the angle ADB, for they are in the same segment, ABDC. And the angle ACB will be equal to the angle ADB, because they are in the same segment, ABCD: therefore the whole angle BDC is equal to the angles ABC, ACB. Take away the angle BAC, which is common: the angles BAC, ABC, ACB, are equal to the angles BAC, BDC. But the angles BAC, ABC, ACB, are equal to two right angles: wherefore also the angles BAC, BDC, are equal to two right angles. In like manner we can demonstrate that the angles ABD, ACD, are also equal to two right angles. Therefore the opposite angles, &c. Q. E. D.

Deduction.

If two opposite angles of any trapezium be equal to two right angles, the other two angles are equal to two right angles, and a circle may be described about it.

PROPOSITION XXIII.

Theorem.

Upon the same straight line, and on the same side of it, two similar segments of circles cannot be described which do not coincide with each other.

For, if it be possible, on the same right line, AB, let two similar segments of circles, ACB, ADB, be described, which do not coincide with each other. Let the circumferences ACB, ADB, meet one another at the points A, B, and they have[a] no other points common except A, B. But between A, B, one will be interior, and the other exterior. In the interior take any point C, and join AC, which, produced, will meet the exterior in D, and join CB, BD. Therefore because ACB is a segment similar to the segment ADB, and segments of circles are similar which contain equal angles,[b] the angle ACB will be equal to the angle ADB, the exterior to the interior, which is impossible. Therefore upon the same right line, &c. Q. E. D.

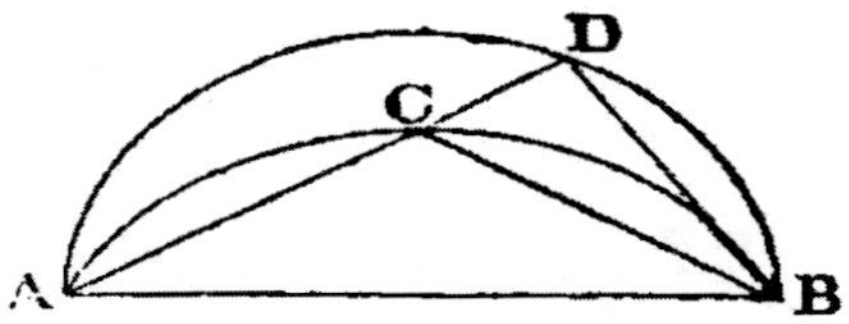

[a] 10. 3.

[b] Def. 10. 3.

Deduction.

Of unequal segments described upon the same base and towards the same parts, the circumference of that which contains the greater angle will be the interior.

PROPOSITION XXIV.

THEOREM.

Similar segments of circles, upon equal right lines, are equal to one another.

For let the similar segments of circles AEB, CFD, be upon the equal right lines AB, CD. The segment AEB is equal to the segment CFD. For the segment AEB coinciding with the segment CFD, and the point A with the point C; but the right line AB, with the right line CD, the point B shall also coincide with the point D, because AB is equal to CD; but AB coinciding with CD, the segment AEB shall coincide with the segment CFD. For if the right line AB coinciding with the right line CD, the segment AEB does not coincide with the segment CFD, it will fall either within or without it, which is impossible.[a] Therefore the right line AB coinciding with the right line CD, the segment AEB cannot but coincide with the segment CFD, and as it coincides with it, it is consequently equal. Therefore similar segments, &c. Q. E. D.

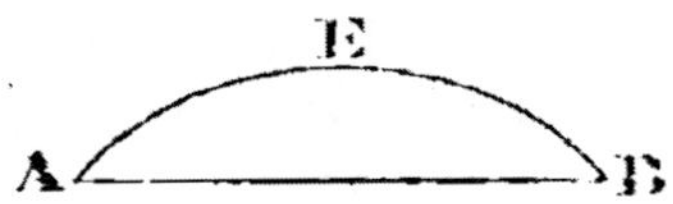

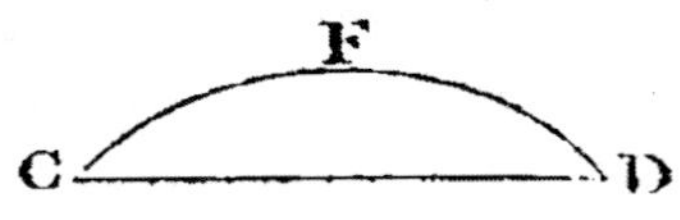

[a] 23. 3.

PROPOSITION XXV.

Problem.

The segments of a circle being given to describe the circle of which it is a segment.

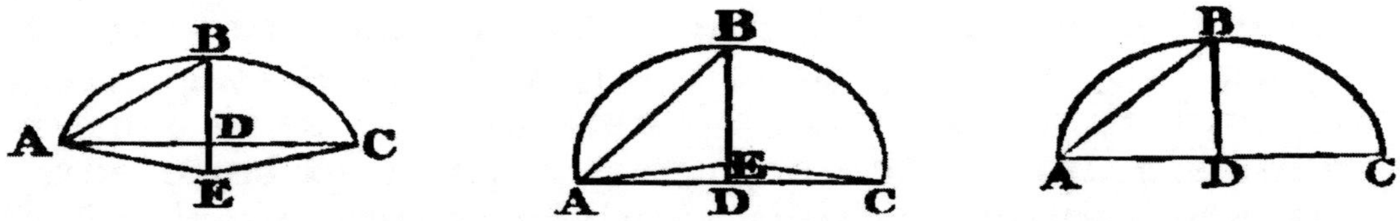

Let ABC be a given segment of a circle. It is required to describe the circle of which ABC is a segment.
a 10. 1. Bisect AC[a] in D, and from the point D draw AC at right
b 10. 1. angles to DB,[b] and join AB. Therefore the angle ABD is either greater than the angle BAD, or less, or equal to it. Let it be greater, and at the right line AB draw
c 23. 1. EC, and at the given point A in it, make[c] the angle BAE equal to the angle ABD; but BD, AE, being produced, will meet one another in E, and join EC. Therefore because the angle ABE is equal to the angle BAE, the
d 6. 1. right line BE[d] will also be equal to AE, and DE is common: the two AD, DE, are equal to the two CD, DE, each to each, and the angle ADE to the angle CDE, for each of them is a right angle. Wherefore also the base
e 4. 1. BE is equal to the base EC.[e] But AE has been shown to be equal to EB: wherefore also BE is equal to EC, and consequently the three right lines AE, EB, EC, are equal to one another. Therefore from the centre E, with the distance of any of them, AE, EB, EC, the circle
f 9. 3. so described will pass through the remaining[f] points, and it will be the circle to be described. Wherefore the segment of a circle being given, the circle of which it is given is described. But it is also evident that the segment ABC is less than a semicircle, because its centre falls without it. In like manner the angle ABD is also equal to the angle BAD, the right line AD will be equal to each of the right lines BD, DC. Therefore the three right lines AD, DB, DC, will be equal to one another, because D will be the centre of the circle described, and the segment ABC a semicircle. But if the angle ABD be less than the angle BAD, describe at the right line BA, and at the given point A in it, the angle ABD equal to the angle BAE, within the segment ABC, the centre E will be in DB, and the segment ABC will be greater

than a semicircle. Therefore the segment of a circle being given, the circle is described of which it is a segment. Q. E. F.

PROPOSITION XXVI.*

THEOREM.

In equal circles equal angles stand upon equal circumferences, whether they stand at the centre or the circumference.

Let ABC, DEF, be equal circles, and BGC, EHF, equal angles in them at the centre; also BAC, EDF, at the circumference. The circumference BAC is equal to the circumference EDF, for join BC, EF. Because the circles ABC, DEF, are equal, the right[a] lines drawn from their centres shall also be equal: therefore the two BG, GC, are equal to the two EH, HF, and the angle at G equal to the angle at H. Wherefore also the base BC[b] is equal to the base EF. Again, because the angle at A is equal to the angle at D, the segment BAC will be similar[c] to the segment EDF, and they are upon equal right lines BC, EF. But similar segments[d] standing upon equal bases have equal circumferences. Therefore the segment BAC is equal to the segment EDF. But the whole ABC is equal to the whole EDF:[e] therefore the remaining segment BKC is equal to the remaining segment ELF. Therefore in equal circles, &c. Q. E. D.

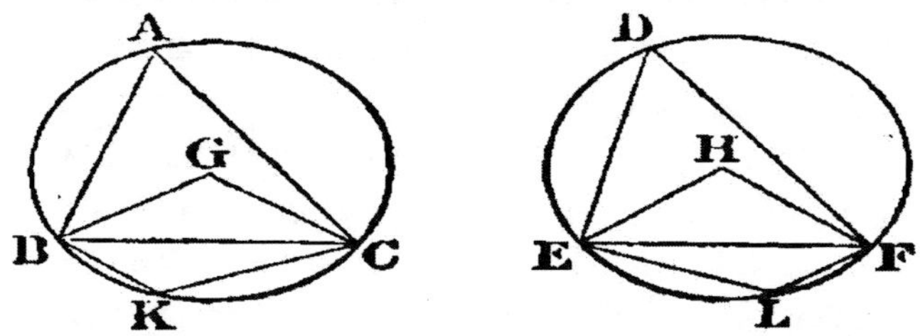

[a] Def. 1.
[b] 4. 1.
[c] Def. 11.
[d] Cor. 24. 3.
[e] Ex hyp.

Deductions.

1. If two equal circles cut each other, and from either point of intersection a line be drawn meeting the circumferences, the part of it intercepted between the circumferences will be bisected by the circle whose diameter is the common chord of the equal circles.

2. The arcs of circles intercepted between two parallel chords are equal to one another.

3. In equal circles the greater angle stands upon the greater circumference.

---

* This and the three succeeding propositions will hold good, if in the same circle.

4. In equal circles, two equal right lines terminated in a point in the circumference of the one being equal to two other right lines terminated in a point in the circumference of the other, then shall the intercepted arcs be equal to one another.

PROPOSITION XXVII.

Theorem.

In equal circles angles which stand upon equal circumferences are equal to one another, whether they stand at the centre or the circumference.

For in the equal circles ABC, DEF, let the angles BGC, EHF, at the centre, also BAC, EDF, at the circumference, stand upon equal circumferences BC, EF. The angle BGC is equal to the angle EHF, and the angle BAC to the angle EDF. If the angles BGC, EHF, be not equal, one of them will be greater than the other. Let BGC be the greater, and make at the right
line BG, and at the point
G in it, the angle BGK
a 23. 1. equal to the angle EHF.[a]
But equal angles stand
upon equal circumferences when they are at
b 26. 3. the centre,[b] wherefore the
circumference BK is equal to the circumference EF. But the circumference EF is equal to the circumference BC: wherefore also BK is equal to BC, the less to the greater, which is impossible. Therefore the angle BGC is not unequal to the angle EHF; wherefore it is equal to it. But the angle which is at A is half of the angle BGC; also the angle at D is half of the angle EHF: wherefore the angle which is at A is equal to the angle
c 20. 3. which is at D.[c] Therefore in equal circles, &c. Q. E. D.

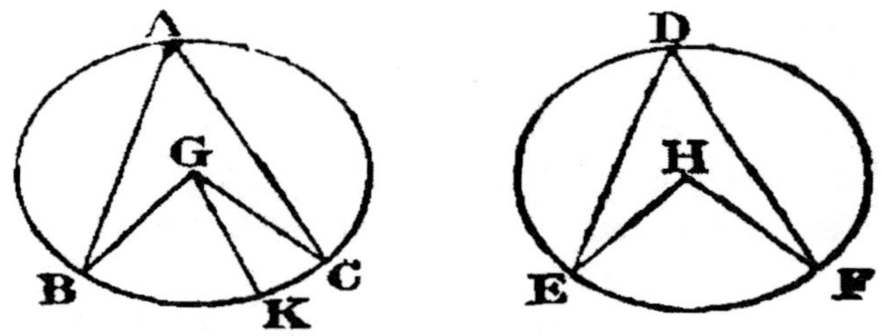

Deduction.

In equal circles, the greater of two circumferences subtends the greater angle, whether those angles be at the centre or the circumference.

PROPOSITION XXVIII.

THEOREM.

In equal circles equal right lines cut off equal circumferences, the greater equal to the greater, and the less to the less.

Let ABC, DEF, be equal circles, and BC, EF, equal right lines in them which cut off the greater circumferences BAC, EDF; also the less circumferences, BGC, EHF. The greater circumference, BAC, is equal to the greater, EDF; and the less circumference, BGC, to the less, EHF.

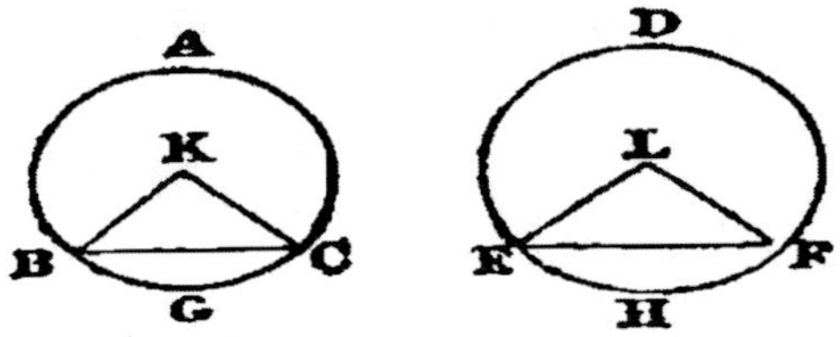

For take K, L, the centres of the circles,[a] and join BK, KC, EL, LF. Because they are equal circles, the right lines drawn from their centres shall be equal: therefore the two BK, KC, are equal to the two EL, LF; and the base BC is equal to the base EF: wherefore the angle BKC is equal to the angle ELF.[b] But equal angles stand upon equal circumferences: wherefore the circumference BGC is equal to the circumference EHF.[c] But the whole circle ABC is equal to the whole circle DEF: therefore the remaining circumference, BAC, will be equal to the remaining circumference, EDF. Wherefore in equal circles, &c. Q. E. D.

[a] 1. 3
[b] 8. 1.
[c] 26. 3.

Deduction.

In equal circles, the greater of two chords cuts off the greater circumference.

PROPOSITION XXIX.

THEOREM.

In equal circles the right lines are equal which subtend equal circumferences.

Let ABC, DEF, be equal circles, and in them take the equal circumferences, BGC, EHF, and join BC, EF. The right line BC is equal to the right line EF. For

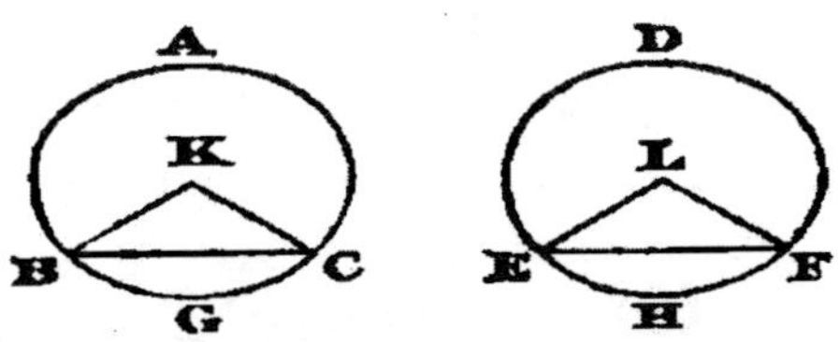

find the centres K, L, of the circles,[a] and join BK, KC, EL, LF. Because, therefore, the circumference BGC is equal to the circumference EHF, the angle BKC will

[a] 1. 3.

also be equal to the angle ELF, and because the circles ABC, DEF, are equal, the right lines drawn from their centres will also be equal; therefore the two BK, KC, are equal to the two EL, LF, and they contain equal
[b] 4. 1. angles; wherefore the base BC is equal to the base EF.[b] Therefore in equal circles, &c. Q. E. D.

Deduction.

In equal circles, the greater of two circumferences is subtended by the greater chord.

PROPOSITION XXX.

Problem.

To bisect a given circumference.

Let ADB be a given circumference. It is required to bisect it. Join AB, and bisect it in C. Also from the point C draw CD at right angles to AB, and join AD, DB. Therefore because AC is equal to CB, also CD is common; the two AC, CD, are equal to the two BC, CD, and the angle ACD is equal to the angle BCD, for each of them is a right angle; wherefore the base AD is equal to the base BD. But equal right lines cut off equal circumferences. Wherefore the circumference AD will be equal to the circumference BD. Therefore the given circumference has been bisected. Q. E. F.

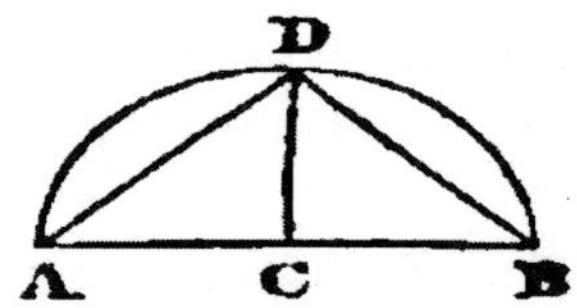

PROPOSITION XXXI.

Theorem.

In a circle the angle in a semicircle is a right angle, also that in a segment greater than a semicircle is less than a right angle, and that in a segment less than a semicircle is greater than a right angle, and moreover the angle of a greater segment is greater than a right angle, but that of a less segment is less than a right angle.

Let ABCD be a circle whose diameter is BC, and E the centre; join BA, AC, AD, DC. The angle which is in the semicircle BAC is a right angle, also that which is in the segment ABC greater than a semicircle; viz. the angle ABC is less than a right angle, and that which is in the segment ADC, which is less than a semicircle, viz. the angle ADC, is greater than a right

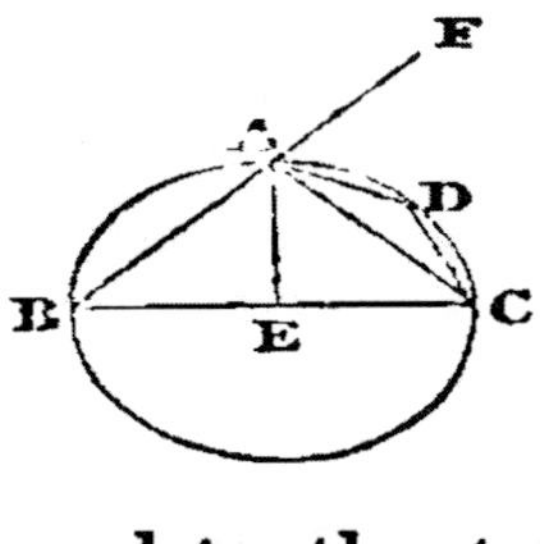

angle. Join AE, and produce BA to F. Therefore because BE is equal to EA, the angle EAB will also be equal to the angle EBA.[a] Again, because AE is equal to EC, the angle ACE will be equal to the angle CAE; therefore the whole angle BAC is equal to the two angles ABC, ACB. But the angle FAC is without the triangle ABC, and is equal to the two ABC, ACB;[b] therefore the angle BAC is equal to the angle FAC; and consequently each of them is a right angle. Wherefore in the semicircle BAC, the angle BAC is a right angle. And because the two angles ABC, BAC, of the triangle ABC, are less than two right angles,[c] but BAC is a right angle; the angle ABC will be less than a right angle, and it is the angle in the segment ABC which is greater than a semicircle. But since ABCD is a quadrilateral figure inscribed in a circle, also the opposite angles of quadrilateral figures are equal to two right angles, the angles ABC, ADC, will be equal to two right angles, and the angle ABC is less than a right angle; therefore the remainder ADC will be greater than a right angle, and it is in the segment ADC, which is less than a semicircle. Moreover the angle of the greater segment which is contained by the circumference ABC, and the right line AC, is greater than a right angle, but the angle of the less segment contained by the circumference ADC and the right line AC, is less than a right angle. Whence it is evident, because the angle which is contained by the right lines BA, AC, is a right angle, that which is contained by the circumference ABC and the right line AC will also be greater than a right angle. Again, because the angle contained by the right lines CA, AF, is a right angle, that which is contained by the right line CA and the circumference ADC, is less than a right angle. Therefore in a circle, &c. Q. E. D.

[a] 5. 1.
[b] 16. 1.
[c] 17. 1.

Deductions.

1. In a right angled triangle, given the hypothenuse and perpendicular let fall from the right angle to the hypothenuse to construct the triangle.

2. If the chords of two arcs of the same circle cut each other at right angles, the squares of the four segments of the chords are, together, equal to the square of the diameter.

3. If the diameter of a circle be divided into any two parts, and from the point of section a perpendicular be drawn to the circumference, the squares of the two parts, with twice the square of the perpendicular, shall be together equal to the square of the diameter.

PROPOSITION XXXII.

Theorem.

If a right line touches a circle, and from the point of contact another right line be drawn cutting the circle, the angles which this line makes with the touching line will be equal to those which are in the alternate segments of the circle.

For let any right line EF touch the circle ABCD in B, and from the point B draw the right line BD anyhow, cutting the circle ABCD. The angles which BD makes with the touching line EF are equal to those which are in the alternate segments of the circle; that is, the angle FBD is equal to the angle which is in the segment DAB; also the angle DBE is equal to the angle in the segment DCB. For from the point B draw BA at right angles to EF, and in the circumference BD take any point C, and join AD, DC, CB. Therefore because any right line EF touches the circle ABCD, in the point B, and from the point of contact B a right line BA is drawn at right angles to the touching line, the centre of the circle ABCD will be in

a 19. 3. BA.[a] Wherefore BA is a diameter of the same circle, and ADB an angle in a semicircle is a right angle. Therefore the remaining angles BAD, ABD, are equal to a right angle. But ABF is also a right angle; wherefore the angle ABF is equal to the angles BAD, ABD. Take away the common angle ABD. Therefore the remainder DBF is equal to that which is in the alternate segment of the circle; namely, the angle BAD. And because ABCD is a quadrilateral figure inscribed in a circle, and its opposite angles are equal

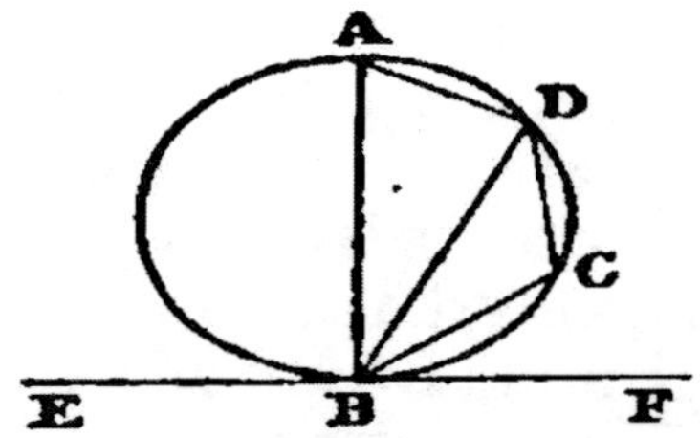

b 22. 3. to two right angles,[b] the angles DBF, DBE, are equal to the angles BAD, BCD, of which DBF is shown to be equal to BAD. Wherefore the remainder DBE will be equal to DCB, viz. to that which is in the alternate segment of the circle DCB. If, therefore, any right line, &c. Q. E. D.

Deductions.

1. If a right line be drawn a tangent to an arc at the point of bisection, it shall be parallel to the chord of the arc.

2. If a triangle be described in a circle, and from the vertex a line be drawn touching the circle, the angles formed by this line, and the two sides of the triangle, shall be respectively equal to the three angles of the triangle.

PROPOSITION XXXIII.

PROBLEM.

Upon a given right line to describe a segment of a circle which shall contain an angle equal to a given rectilineal angle.

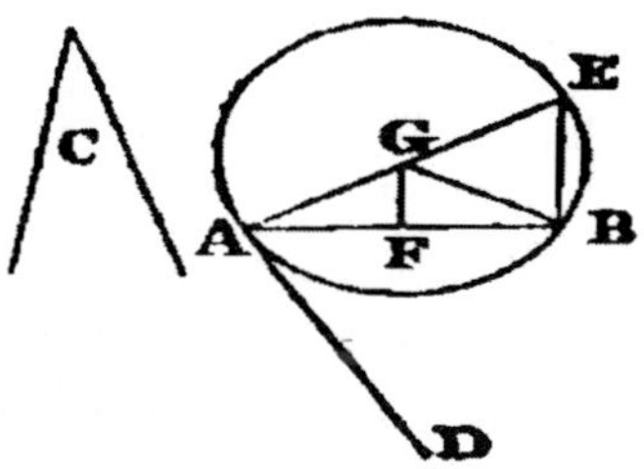

Let AB be a given right line, and C a given rectilineal angle. It is required upon the given right line AB to describe a segment of a circle which shall contain an angle equal to the angle at C. At the right line AB, and at the given point A in it, make the angle BAD[a] equal to the angle C, and from the point A draw AE at right angles to AD. But bisect AB in F, and from the point F draw FG at right angles to AB, and join GB. Therefore because AF is equal to FB, and FG common, the two AF, FG, are equal to the two BF, FG, and the angle AFG to the angle BFG. Wherefore the base AG is equal to the base GB. Therefore from the centre G with the distance AG, the circle described will pass through B. Let it be described, and let it be AKE. Therefore because from the extremity of the diameter AE, and from the point A, AD is drawn at right angles to AE, AD shall touch the circle. And because a certain right line AD touches the circle ABE, and from the point of contact, A, a right line AB is drawn into the circle ABE, the angle DAB will be equal to that in the alternate segment of the circle, viz. to AEB. But the angle DAB is equal to the angle C. Wherefore also the angle C will be equal to the angle AEB. Therefore upon a given right line AB, a segment of a circle AEB has been described containing an angle AEB, equal to the given angle at C. Q. E. F.

[a] 23. 1.

Deductions.

1. Upon a given finite right line to describe the segment of a circle, which shall be similar to a given segment.

2. Given the base, the verticle angle, and the difference of the other two sides, to construct the triangle.

3. The base, the vertical angle, and the altitude, being given to construct the triangle.

PROPOSITION XXXIV.

Problem.

From a given circle to cut off a segment which shall contain an angle equal to a given rectilineal angle.

Let ABC be a given circle, and D the given rectilineal angle. It is required from the circle ABC to cut off a segment which shall contain an angle equal to the angle D. Draw the right line EF, touching the circle ABC, in the point B, and at the right line BF, and at the point B in it make the angle FBC equal to the angle D. Therefore because a certain right line EF touches the circle ACB in the point B, and from the point of contact, BC is drawn, the angle FBC will be equal to that in the alternate segment of the circle BAC.[a] But the angle FBC is equal to the angle at D; wherefore also the angle in the segment BAC will be equal to the angle at D. Therefore from a given circle ABC, a segment BAC is cut off containing an angle equal to the given rectilineal angle at D. Q. E. F.

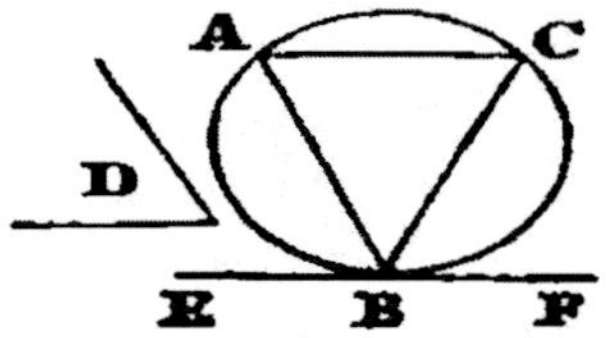

[a] 32. 3.

Deduction.

From two circles cut off two similar segments.

PROPOSITION XXXV.

Theorem.

If in a circle two right lines mutually cut one another, the rectangle contained under the segments of one of them is equal to the rectangle contained under the segments of the other.

For in the circle ABCD let the two right lines AC, BD, mutually cut one another in the point E. The rectangle

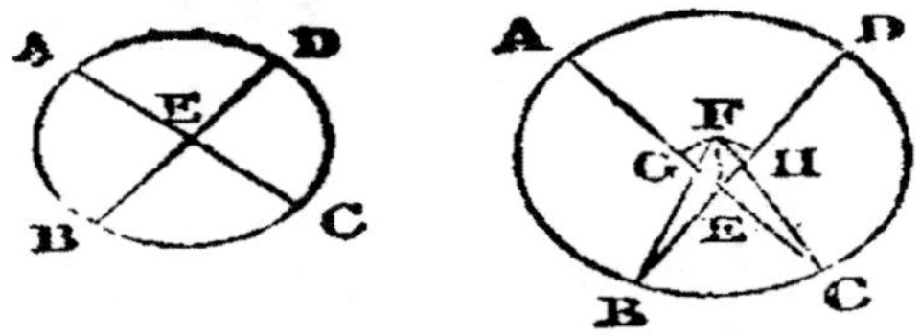

contained under AB, EC, is equal to that contained under DE, EB. If AC, BD, pass through the centre so that E be the centre of the circle ABCD, it is manifest the right lines AE, EC, DE, EB, being equal, the rectangle contained under AE, EC, is equal to that which is contained under DE, EB. If AC, DB, do not pass through the centre, find the centre of the circle ABCD, which let be F, and from F draw FG, FH, perpendicular to AC, BD, and join FB, FC, FE. Because therefore a certain right line GF drawn through the centre cuts the right line AC not drawn through the centre at right angles, it shall bisect it;[a] wherefore AG [a] 3. 3.
is equal to AC, and because the right line AC is divided into equal parts at the point G, and into unequal at the point E, the rectangle contained under AE, EC, together with the square of EG, will be equal to the square of GC,[b] add the common square of GF. Wherefore the [b] 5. 2.
rectangle contained under AE, EC, together with the squares of EG, GF, is equal to the squares CG, GF, but the square of FE is equal to the squares of EG, GF;[c] also [c] 47. 1.
the square described upon FC is equal to the squares of CG, GF. Therefore the rectangle under AE, EC, together with the square of FE, is equal to the square of FC. But CF is equal to FB; wherefore the rectangle under AE, EC, together with the square of EF, is equal to the square described upon FB. For the same reason, the rectangle under DE, EB, together with the square of FE, is equal to the square of FB. But it was shown that the rectangle under AE, EC, together with the square of FE, is equal to the square of FB. Wherefore the rectangle under AE, EC, together with the square of FE, is equal to the rectangle under DE, EB, together with the square of FE; take away the common square of FE; therefore the remaining rectangle under DE, EC, will be equal to the remaining rectangle under DE, EB. Wherefore if in a circle, &c. Q. E. D.

Deduction.

To make a rectangle which shall be equal to a given square, and shall have its two adjacent sides, together, equal to a given right line, the side of the given square being less than half of the given right line.

PROPOSITION XXXVI.

Theorem.

If any point be taken without a circle, and from it two right lines be let fall on the circle, one of which cuts the circle, and the other touches it, the rectangle which is contained by the whole cutting line, and that part between the point taken without the circle, and the convex circumference of the circle, will be equal to the square of the touching line.

For without the circle ABC take any point D, and from it let fall the two right lines DCA, DB, to the said circle, and let DCA cut the circle ABC, but DB touch it. The rectangle contained under AD, DC, is equal to the square of DB. For DCA either passes through the centre, or it does not. First, let DA pass through the centre of the circle ABC,

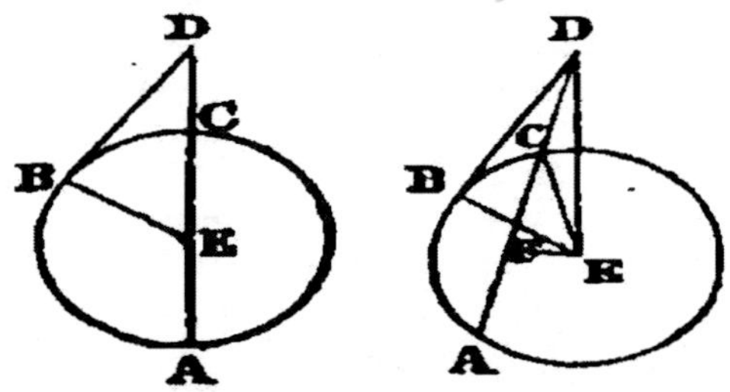

which let be E, and join EB. The angle EBD will be a right angle, because the right line AC is bisected in E, and CD is added to it; the rectangle under AD, DC, together with the square of EC, will be equal to the square of ED, but CE is equal to EB; wherefore the rectangle under AD, DC, together with the square of EB, will be equal to the square of ED, but the square of ED is equal to the squares of EB, BD, for EBD is a right
a 47. 1. angle.[a] Take away the common square of EB; wherefore the remaining rectangle under AD, DC, will be equal to the square of DB. Secondly, let DCA not pass through the centre of the circle ABC, and find the centre E, and draw EF perpendicular to AC, and join EB, EC, ED. Therefore EFD is a right angle. And because a certain right line EF drawn through the centre cuts the right line AC not drawn through the centre at right angles, it shall also bisect it; wherefore AF is equal to FC. Again, because the right line AC is bisected in F, and CD is added to it, the rectangle under AD, DC, together with the square of FC, is equal
a 5. 2. to the square of FD;[a] add the common square of FE; therefore the rectangle under AD, DC, together with the squares of FC, FE, is equal to the squares of DF, FE. But the square of DE is equal to the squares of DF, FE, because

EFD is a right angle, but the square of CE is equal to the squares of CF, FE. Wherefore the rectangle under AD, DC, together with the square of CE, is equal to the square of ED, but CE is equal to EB. Therefore the rectangle under AD, DC, together with the square of EB, is equal to the square of ED. But the squares of EB, BD, are equal to the square of ED, for EBD is a right angle. Wherefore the rectangle under AD, DC, together with the square of EB, is equal to the squares of EB, BD; take away the common square of EB: therefore the remaining rectangle under AD, DC, will be equal to the square of DB. If, therefore, any point, &c. Q. E. D.

COROLLARIES.

1. (Clavius.) From this 36th proposition, it is manifest if, from any point without a circle, several right lines are drawn cutting the circle, the rectangles contained under the whole lines and the parts without the circle, are equal to one another.

2. (Clavius.) It is also proved that two right lines drawn from the same point which touch the circle are equal to one another.

3. (Clavius.) It is also evident from the same point only two right lines can be drawn which can touch the circle.

PROPOSITION XXXVII.

THEOREM.

If any point be taken without a circle, and from it two right lines be let fall to the circle, one of which cuts the circle, and the other falls upon it; also let the rectangle contained by the whole line cutting the circle, and the part between the point taken without the circumference, and the convex circumference be equal to the square of the line meeting the circle, the line which meets it shall touch the circle.

For without the circle ABC take any point D, and from it let fall the two right lines DCA, DB. Let DCA cut the circle, and DB fall upon it, and let the rectangle AD, DC, be equal to the square of DB; then DB touches the circle ABC. For draw the right line DE touching the circle ABC, and find the centre of the circle ABC,[a] [a] 1. 3. which let be F; join FE, FB, FD; wherefore the angle FED is a right angle. And because DE touches the circle ABC, but DCA cuts it, the rectangle under AD, DC, will be equal to the square of DE; but the rectangle under AD, DC, is equal to the square of DB; therefore the square of DE will be equal to the square of DB, and consequently the right line DE will be equal to the right line DB, but FE is equal to FB. Therefore the two DE, EF, are equal to the two DB, BF, and the base FD common; therefore the angle DEF is equal to the angle DBF. But DEF is a right angle; wherefore also DBF is a right angle, and the diameter FB is drawn. But the right line drawn from the extremity of the diameter of

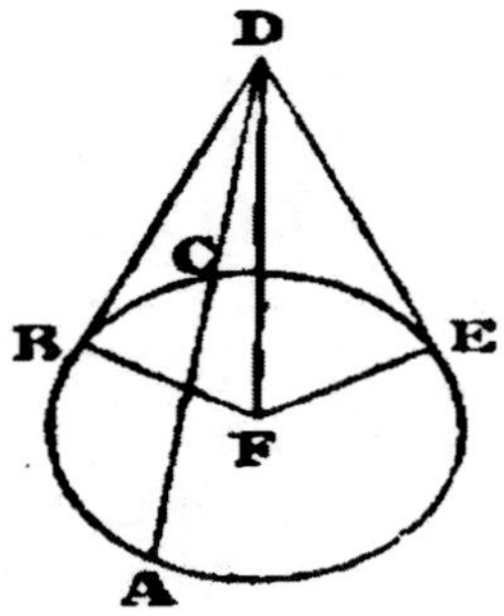

b 21. 3. a circle at right angles touches the circle;[b] wherefore DB must touch the circle ABC. If, therefore, any point, &c. Q. E. D.

Deductions.

1. To describe a circle which shall touch two given right lines and pass through a given point between them.

2. To describe a circle which shall pass through two given points, and touch a given right line, the given points being both on the same side of the right line.

EUCLID'S ELEMENTS.

BOOK IV.

DEFINITIONS.

1. A rectilineal figure is said to be inscribed in a rectilineal figure, when every angle of the inscribed figure touches every side of the figure in which it is inscribed.*

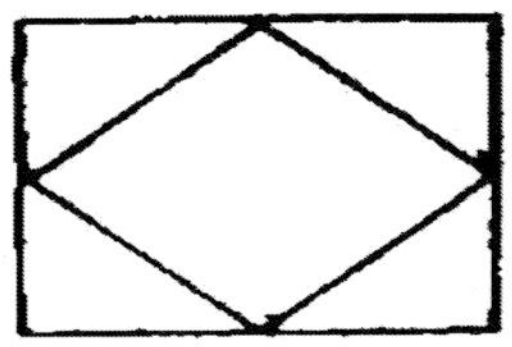

2. In like manner a figure is said to be circumscribed about a figure, when all the sides of the circumscribed figure touch all the angles of that figure about which it is circumscribed.

3. A rectilineal figure is said to be inscribed in a circle, when every angle of the inscribed figure is upon the circumference of the circle.

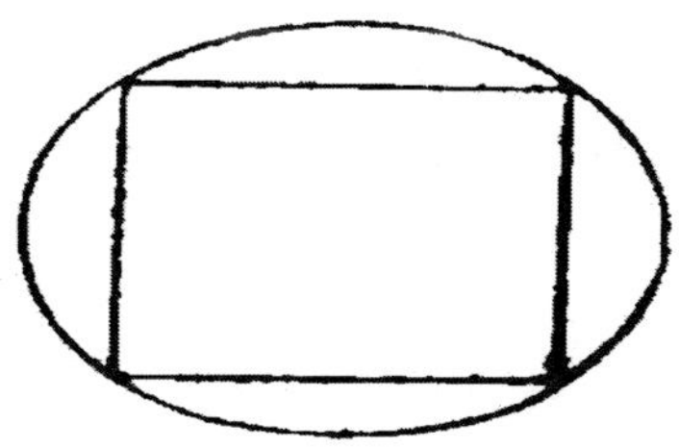

4. A rectilineal figure is said to be described about a circle, when every side of the circumscribed figure touches the circumference of the circle.

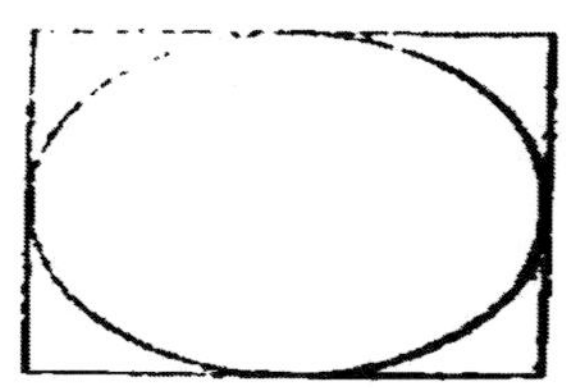

5. In like manner a circle is said to be inscribed in a rectilineal figure, when the circumference of the circle touches every side of the figure in which it is inscribed.

---

* When a figure is within another, so that all the angles of the inner figure are upon the sides of the figure in which it is, this figure is said by the Greeks, ἐγγράφεσθαι, to be inscribed within the other, and the outward figure is said, περιγράφεσθαι, to be circumscribed about the inner one; but when it is merely to describe a circle, as in the 25th prop. lib. 3, προσαναγράφω is used. This distinction is also observed in Ptolemy's Μεγαλη Συνταξις, as may be seen in the ninth chapter of the First Book.

6. A circle is said to be circumscribed about a figure, when the circumference of the circle touches every angle of the figure about which it is circumscribed.

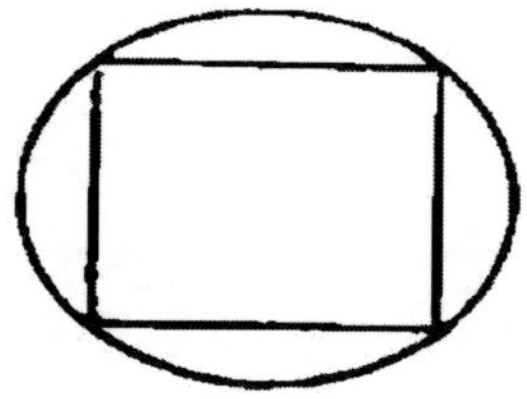

7. A right line is said to be applied in a circle when its extremities are in the circumference of the circle.

PROPOSITION I.

PROBLEM.

In a given circle to apply a right line equal to a given right line not greater than the diameter of the circle.

Let ABC be the given circle, and D the given right line not greater than the diameter of the circle; it is required in the circle ABC to apply a right line equal to the right line D. Draw BC the diameter of the circle ABC. If, therefore, BC is equal to D, what was proposed will now be done. For in the circle ABC, BC is applied equal to the right line D. But if BC is greater than D, make CE equal to D,[a] and with centre C, and distance CE, describe the circle AEF and join CA. Therefore, because the point C is the centre of the circle AEF, CA is equal to CE.[b] But CE is equal to D; therefore D also is equal to CA. Therefore in the given circle ABC, &c. Q. E. F.

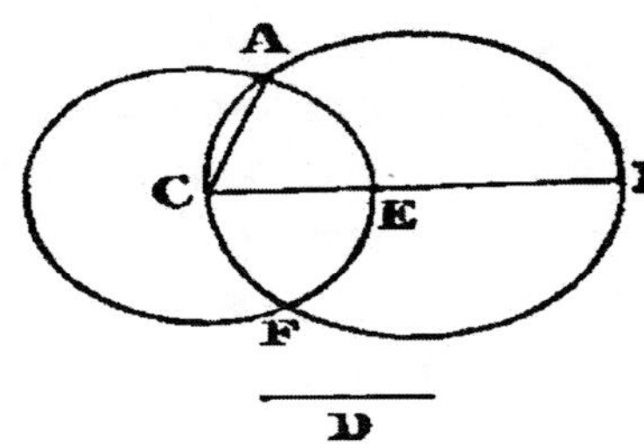

[a] 3. 1.

[b] 15 Def. 1.

PROPOSITION II.

PROBLEM.

In a given circle to inscribe a triangle equiangular to a given triangle.

Let ABC be the given circle, and DEF the given triangle; it is required in the circle ABC to inscribe a triangle equiangular to the triangle DEF.

Draw GH touching the circle ABC in the point A, and at the right line GH, and at the point A in it, make the angle HAC equal to the angle DEF;[a] again, at the right line GA, and at the point A in it, make the angle GAB equal to the angle FDE, and join BC.

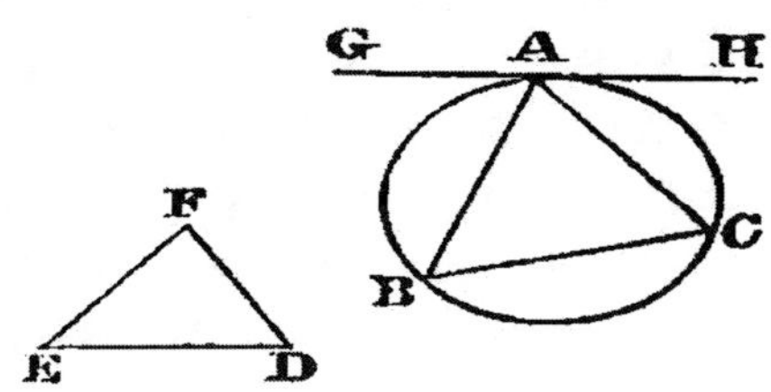

Therefore, because some right line HG touches the circle ABC, and from the point of contact at A, AC is drawn in the circle, whence the angle HAC is equal to the angle ABC[b] in the alternate segment of the circle. But the angle HAC is equal to DEF. For the same reason, the angle ACB is equal to the angle FDE, and

[a] 23. 1.

[b] 32. 3.

therefore the remaining angle BAC is equal to the re-
c 32. 1. maining angle EFD.[c] Therefore the triangle ABC is equiangular to the triangle DEF, and is inscribed in
d 3 Def. 4. the circle ABC.[d] Wherefore in a given circle, &c. Q. E. F.

PROPOSITION III.

Problem.

About a given circle to circumscribe a triangle equiangular to a given triangle.

Let ABC be a given circle, and DEF a given triangle; it is required about the circle ABC to circumscribe a triangle equiangular to the given triangle DEF.

Produce EF both ways to the points G, H, and find
a 1. 3. K the centre of the circle ABC,[a] also draw anyhow the right line KB; and at the right line KB, and at the point K in it, make the angle BKA equal to the
b 23. 1. angle DEG,[b] also the angle BKC equal to DFH, and through the points A, B, C, draw the right lines LAM, MBN, NCL, touching
c 17. 3. the circle ABC.[c]

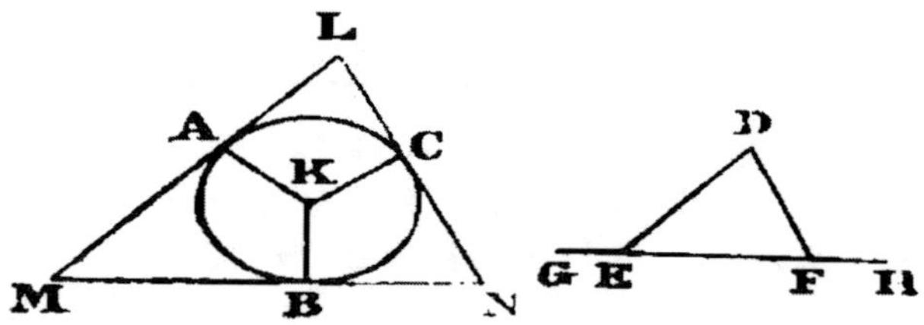

And because the lines LM, MN, NL, touch the circle ABC in the points A, B, C, and KA, KB, KC, are joined;
d 18. 3. the angles at the points A, B, C, are right angles.[d] And because the four angles of the quadrilateral figure
e 32. 1. AMBK are equal to four right angles,[e] for it may be divided into two triangles, of which the angles MAK, KBM, are right ones; therefore the remaining angles AMB, AKB, are equal to two right ones; but DEG,
f 13. 1. DEF, are also equal to two right angles;[f] therefore AKB, AMB, are equal to DEG, DEF, of which AKB is equal to DEG; whence the remaining angle AMB is equal to the remaining angle DEF. In like manner, it may be shown that LNM is equal to DEF; and therefore the remaining angle MLN is also equal to the remaining angle EDF. Therefore LMN is a triangle equiangular to the triangle DEF. Therefore about a given circle, &c. Q. E. F.

PROPOSITION IV.

PROBLEM.

To inscribe a circle in a given triangle.

Let ABC be the given triangle; it is required to inscribe a circle in the triangle ABC.

Bisect the angles ABC, ACB, by the right lines BD, CD,[a] and let them meet one another in the point D, [a] 9. 1. and draw from the point D to AB, BC, CA, the perpendicular right lines DE, DF, DG.[b] [b] 12. 1.

And because the angle ABD is equal to the angle DBC, and BED is a right angle, consequently equal to the right angle BFD; therefore EBD, FBD, are two triangles, having two angles equal to two angles, and one side to one side; viz. the side BD opposite to one of the equal angles common to them both, and, therefore, the remaining sides of the one shall be equal to the remaining sides of the other;[c] whence DE is equal [c] 26. 1. to DF. For the same reason, DG is equal to DF. Therefore the three right lines DE, DF, DG, are equal to one another; wherefore from the centre D, and with the distance any one of them DE, DF, DG, the circle described will pass through the remaining points, and will touch the right lines AB, BC, CA, wherefore the angles at the points E, F, G, are right ones. For if it cut them, a line drawn at right angles to the diameter of the circle from the extremity will fall within the circle, which has been shown to be absurd.[d] [d] 16. 3. Therefore, with the centre D, and distance any one of them DE, DF, DG, the circle described will not cut the right lines AB, BC, CA; whence it touches them, and the circle will be inscribed in the triangle ABC.[e] [e] 5 Def. 4. Therefore in a given triangle, &c. Q. E. F.

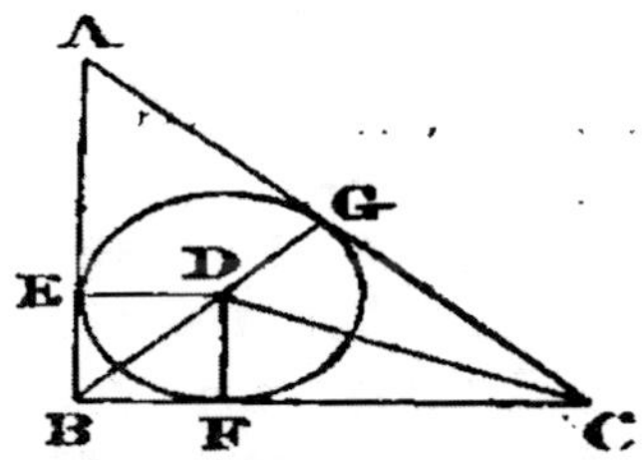

Deductions.

1. The three right lines which bisect the three angles of a triangle meet in the same point.
2. If a circle be inscribed in a right angled triangle, the excess of the two sides containing the right angle above the third side is equal to the diameter of the inscribed circle.

3. In an isosceles triangle, the perpendicular drawn from the vertex bisecting the base passes through the centre of the inscribed circle.

4. In a given circle, to inscribe three equal circles touching each other and the given circle.*

PROPOSITION V.

Problem.

To circumscribe a circle about a given triangle.

Let ABC be the given triangle; it is required to circumscribe a circle about the given triangle ABC.
[a] 10. 1. Bisect the right lines AB, AC, in the points D, E,[a] and

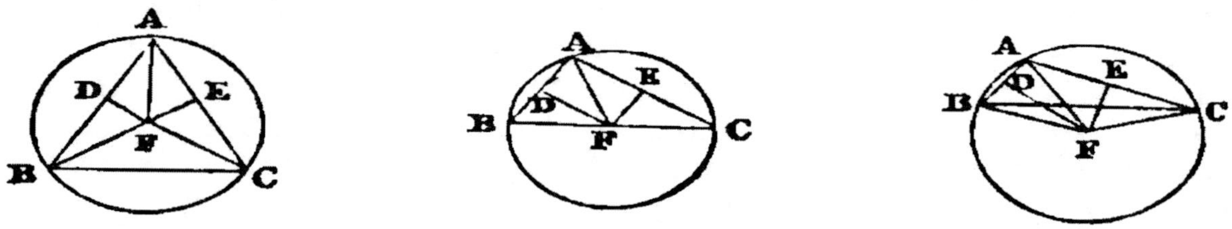

from the points D, E, draw DF, FE, at right angles to
[b] 11. 1. AB, AC.[b] They will meet either within the triangle ABC, or in the right line BC, or without BC.

First, therefore, let them meet within the triangle in F, and join FB, FC, FA. And because AD is equal to BD, and the common side DF is at right angles; therefore the
[c] 4. 1. base AF is equal to FB.[c] In like manner, we show also that CF is equal to AF, wherefore FB is also equal to FC; therefore the three FA, FB, FC, are equal to one another. Therefore with centre F, and distance any one of them FA, FB, FC, a circle described will pass through the remaining points, and the circle will be circumscribed
[d] 6 Def. 4. about the triangle ABC.[d] Let it be circumscribed as ABC.

But also, secondly, let DF, EF, meet in the right line BC in F, as it does in the second figure, and join AF. In like manner it may be proved, that the point F is the centre of the circle circumscribed about the triangle ABC.

But, thirdly, let DF, EF, meet without the triangle ABC in F, as in the third figure, and join AF, BF, CF. And because AD is equal to DB, also the common side DF

---

* The student may find a very neat solution of this problem in Mr. Thomas Simpson's Algebra.

is at right angles, therefore the base AF is equal to FB.[c] [c] 4. 1.
In like manner we show also that FC is equal to FA, wherefore FB is equal to FC; therefore again with centre F, and distance any one of them, FA, FB, FC, a circle described will pass through the remaining points, and it will be circumscribed about the triangle ABC. Let it be described as ABC. Therefore a circle, &c. Q. E. F.

COROLLARY.

Hence it is manifest, when the centre of the circle falls within the triangle, the angle BAC, in a segment greater than a semicircle, is less than a right angle; but when the centre falls on the right line BC, the angle BAC, in a semircircle, is a right angle; and when the centre of the circle falls without the triangle BAC, in a segment less than a semicircle, is greater than a right angle. Wherefore, also, when the given angle is less than a right angle, AF, EF, will meet within the triangle, but when it is a right angle, on BC, and when a greater than a right angle, without BC.

Deduction.

In an equilateral triangle, the centre of the circumscribed circle coincides with the centre of the inscribed circle.

PROPOSITION VI.

PROBLEM.

To inscribe a square in a given circle.

Let ABCD be the given circle; it is required to inscribe a square in the circle ABCD.

Draw the two diameters AC, BD, of the circle ABCD at right angles to one another, and join AB, BC, CD, DA.

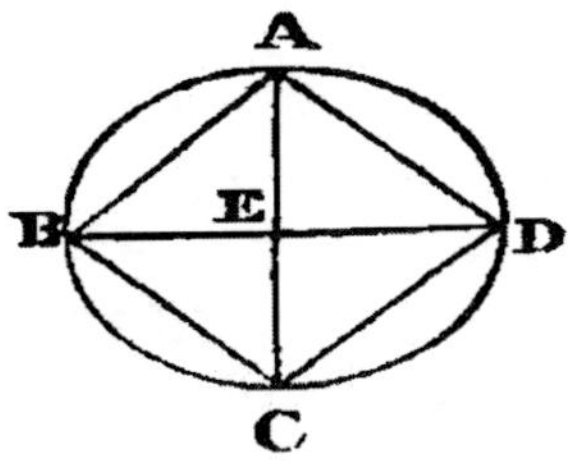

And because BE is equal to ED, for E is the centre, and EA common, is at right angles; therefore the base AB is equal to the base AD;[a] and for the same reason BC, CD, are each equal to BA, AD; therefore the quadrilateral figure ABCD is equilateral. It is also rectangular. [a] 4. 1.

For because the right line BD is a diameter of the circle ABCD, therefore BAD is a semicircle; consequently BAD is a right angle;[b] for the same reason [b] 31. 3.

each of the angles ABC, BCD, CDA, is a right angle; therefore the quadrilateral figure ABCD is rectangular; and it has been shown to be equilateral; wherefore it is a square, and is inscribed in the given circle ABCD. Therefore in a given circle, &c. Q. E. I

Deductions.

1. The square inscribed in a circle is equal to twice the square of half the diameter.

2. To inscribe in a given circle a rectangle, which shall be equal to a given rectangle, whose diameter is equal to the diameter of the given circle.

PROPOSITION VII.

Problem.

To circumscribe a square about a given circle.

Let ABCD be a given circle; it is required to circumscribe a square about the circle ABCD.

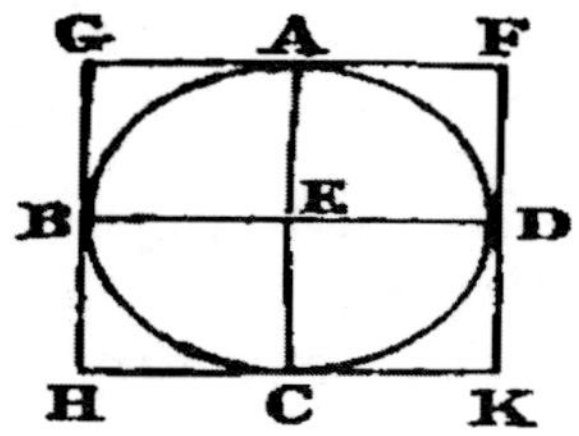

Draw the two diameters AC, BD of the circle ABCD at right angles to one another, and through the points A, B, C, D, draw FG, GH, HK, KF, touching
a 17. 1. the circle ABCD.[a]

Therefore because FG touches the circle ABCD, and from E, the centre, draw EA to the point of contact A; therefore the
b 18. 3. angles at A are right angles.[b] For the same reason the angles at the points B, C, D, are right angles. And because AEB is a right angle, also EBG is a right
c 28. 1. angle; therefore GH is parallel to AC.[c] For the same reason AC is parallel to FK; wherefore also GH is parallel to FK. In like manner we show that each of the lines GF, HK, is parallel to BED. Therefore GK, GC, AK, FB, BK, are parallelograms; whence GF is equal
d 34. 1. to HK,[d] also GH to FK. And because AC is equal to BD, but also AC is equal to each of the lines GH, FK, and ED to each of the lines GF, HK, therefore GH, FK are equal to GF, HK, each to each. Whence the quadrilateral figure FGHK is equilateral. It is also rectangular. For because GBEA is a parallelogram, and AEB is a right angle, therefore AGB is a right angle. In like manner, we show that the angles at H, K, F, are right angles; therefore the quadrilateral figure FGHK is rectangular; and it has been shown to be equilateral;

therefore it is a square. And it is circumscribed about the circle ABCD. Therefore about a given circle, &c. Q. E. F.*

PROPOSITION VIII.

PROBLEM.

To inscribe a circle in a given square.

Let ABCD be a given square; it is required to inscribe a circle in the square ABCD.

Bisect each of the lines AB, AD, in the points E, F,[a] [a] 10. 1.
and through E draw EH parallel to either of them AB, CD; and through F draw FK parallel to either of them AD, BC;[b] [b] 31. 1.
therefore each of them KB, AH, HD, AG, GC, BG, GD, is a parallelogram, and their opposite sides are equal.[c] [c] 34. 1.
And because AD is equal to AB, and AE is half of AD, also AF half of AB, therefore AE is equal to AF; and their opposite sides are equal, whence FG is equal to GE. In like manner we show that each of them FG, GE, is equal to each of them GH, GK; therefore the four GE, GF, GH, GK, are equal to one another. Whence with centre G, and distance any one of them GE, GF, GH, GK, a circle described will pass through the remaining points, and touch the right lines AB, BC, CD, DA, wherefore the angles at E, F, H, K, are right angles. For if the circle cut the lines AB, BC, CD, DA, the right line drawn from the extremity at right angles to the diameter of the circle, that right line will fall within the circle, which has been shown to be absurd.[d] [d] 16. 3.
Therefore with the centre G and distance any one of them GE, GF, GH, GK, a circle described will not cut the right lines AB, BC, CD, DA. Therefore it will touch them, and will be inscribed in the square ABCD. Therefore in a given square, &c. Q. E. F.

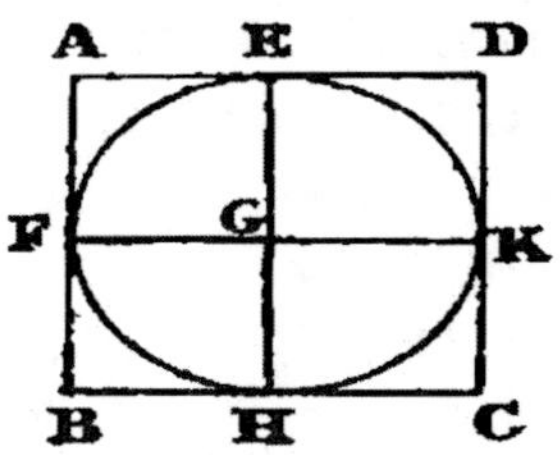

Deduction.

To inscribe a circle in a given rhombus.

---

* If a regular polygon of any number of sides be inscribed in a circle, and it is required to circumscribe about that circle a regular polygon of the same number of sides, and reciprocally, the circumscribed polygon being given to construct the inscribed polygon. See Lacroix's *Elémens de Géométrie*, page 96.

PROPOSITION IX.

Problem.

To circumscribe a circle about a given square.

Let ABCD be a given square; it is required to circumscribe a circle about the square ABCD. For join AC, BD, which will cut one another in the point E.

And because DA is equal to AB, and AC common, the two DA, AC, are equal to the two BA, AC, and the base DC will be equal to the base BC, and the angle
[a] 8. 2. DAC equal to the angle BAC;[a] therefore the angle DAB is bisected by the right line AC. In like manner we demonstrate, that each of the angles ABC, BCD, CDA, is bisected by the right lines AC, DB. And because the angle DAB is equal to the angle ABC, and EAB is the half of DAB, also EBA is the half of ABC; therefore EAB is equal to EBA.

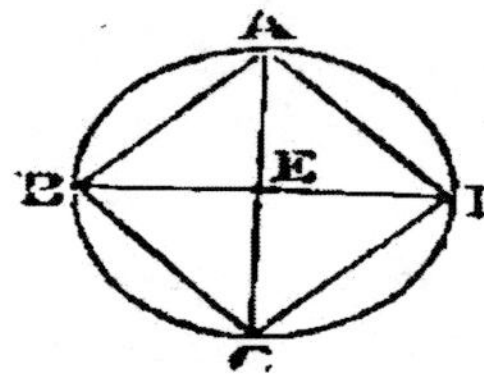

[b] 6. 1. Wherefore also the side EA is equal to the side EB.[b] In like manner we demonstrate that each of the right lines EA, EB, is equal to each of the right lines EC, ED; therefore the four EA, EB, EC, ED, are equal to one another. Whence with centre E and distance any one of them EA, EB, EC, ED, the circle described will pass through the remaining points, and will be circumscribed about the square ABCD. Therefore a circle has been circumscribed, &c. Q. E. F.

Deduction.

To describe a circle about a given parallelogram.

PROPOSITION X.

Problem.

To make an isosceles triangle having each of the angles at the base double of the remaining angle.

Let AB be any given right line, and
[a] 11. 2. cut it in the point C,[a] so that the rectangle contained under AB, BC, may be equal to the square of CA, and with cen-
[b] 3 post. tre A, and distance AB, describe the cir-
[c] 1. 4. cle BDE,[b] also in the circle BDE apply the right line BD[c] equal to the right line AC, not greater than the diameter of the circle BDC; and join AD, CD, also cir-

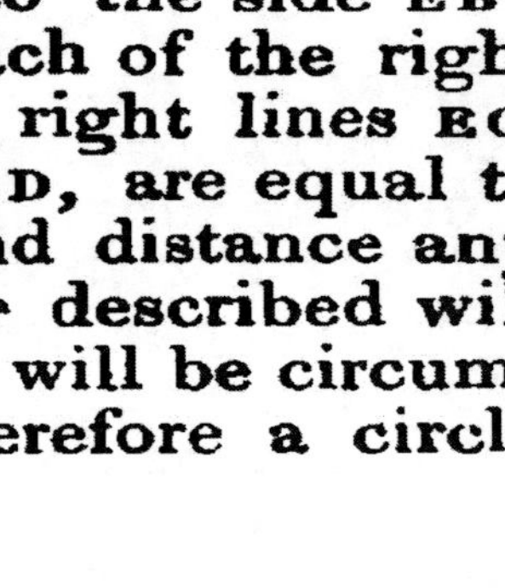

cumscribe the circle ACD about the triangle ACD.[d] d 5. 4.
And because the rectangle contained under AB, BC, is
equal to the square of AC, and AC is equal to BD,
therefore the rectangle under AB, BC, is equal to the
square of BD. Also because some point B is taken
without the circle ACD, and from B let two right lines
fall on the circle ACD, whereof one cuts and the other
touches the circle; also the rectangle under AB, BC, is
equal to the square of BD. Therefore BD touches the
circle ACD.[e] And because BD touches, and from the e 37. 3.
point of contact D, DC is drawn; whence the angle
BDC is equal to the angle DAC,[f] in the alternate seg- f 32. 3.
ment of the circle. Because, therefore, BDC is equal
to DAC, add CDA, which is common. Wherefore the
whole BDA is equal to the two CDA, DAC. But the
exterior angle BCD is equal to CDA, DAC.[g] Therefore g 32. 1.
BDA is equal to BCD. But BDA is equal to CBD, since
the side DA is equal to AB;[h] wherefore also DBA is h 5. 1.
equal to BCD. Therefore the three BDA, DBA, BCD,
are equal to one another. And because the angle DBC
is equal to the angle BCD, the side BD is also equal to
the side DC. But BD is put equal to CA; and AC,
therefore, is equal to CD; wherefore also the angle
CDA is equal to the angle DAC; whence CDA, DAC, are
together double of DAC. And BCD is equal to CDA,
DAC, and BCD is therefore double of DAC. But BCD
is equal to each of them BDA, DBA, and, consequently,
each of them BDA, DBA, is double of BAD. Therefore
an isosceles triangle has been made, &c. Q. E. F.

PROPOSITION XI.

PROBLEM.

To inscribe an equilateral and equiangular pentagon in a given circle.

Let ABCDE be the given circle; it is required in the
circle ABCDE to inscribe an equilateral and equiangular
pentagon. Let FGH be an isosceles triangle, having
each of the angles at G and H double of the angle
at F,[a] and inscribe in the circle ABCDE a 10. 4.
the triangle ACD,[b] equiangular to the b 2. 4.
triangle FGH, so that the angle CAD
may be equal to the angle F, also each
of the angles at G, H, is equal to each
of the angles ACD, CDA; and there-

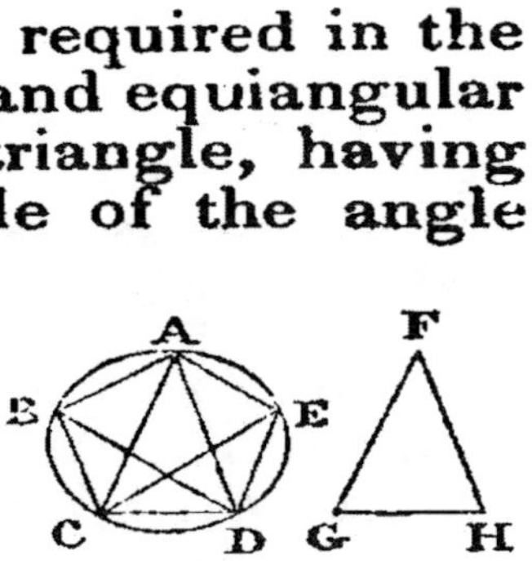

fore each of the angles ACD, CDA, is double of CAD. Bisect each of the angles ACD, CDA, by each of the
c 9. 1. right lines CE, DB,[c] and join AB, BC, DE, EA.

Therefore because each of the angles ACD, CDA, is double of the angle CAD; and are bisected by the right lines CE, DB; therefore the five angles DAC, ACE, ECD, CDB, BDA, are equal to one another, and equal angles
d 26. 2. stand upon equal circumferences;[d] therefore the five circumferences AB, BC, CD, DE, EA, are equal to one another. But equal right lines subtend equal circum-
e 29. 3. ferences;[e] whence the five right lines AB, BC, CD, DE, EA, are equal to one another; therefore the pentagon ABCDE is equilateral. It is also equiangular. For because the circumference AB is equal to the circumference DE, add BCD, which is common; and the whole circumference ABCD is equal to the whole circumference EDCB. And the angle AED stands upon the circumference ABCD, also the angle BAE upon the circumference EDCB, therefore the angle BAE is equal to
f 27. 3. the angle AED.[f] For the same reason each of the angles ABC, BCD, CDE, is equal to each of the angles BAE, AED, therefore the pentagon ABCDE is equiangular. And it has been demonstrated to be equilateral. Therefore in a given circle, &c. Q. E. F.

Deduction.

The angle of a regular pentagon exceeds a right angle by the fifth part of a right angle, and generally if n represent any number of sides greater than four of a regular polygon, each angle will exceed a right one by $(1 - \frac{4}{n})$ of a right angle.

PROPOSITION XII.

Problem.

To circumscribe an equilateral and equiangular pentagon about a given circle.

Let ABCDE be the given circle; it is required to circumscribe an equilateral and equiangular pentagon about the given circle ABCDE.

Let the points of the angles of a pentagon inscribed in the circle be A, B, C, D, E, so that the circumferences AB, BC, CD, DE, EA, be equal,[a] and through A, B, C, D, E, draw GH, HK, KL, LM, MG, touching the circle,[b] and find F the centre of the circle ABCDE, and join FB, FK, FC, FL, FD. [a] 11. 4. [b] 17. 3.

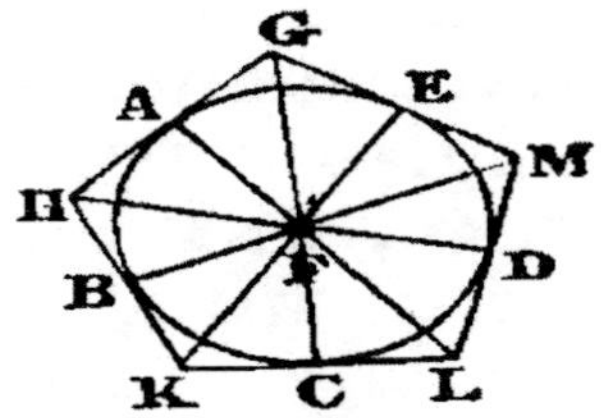

And because the right line KL touches the circle ABCDE in C, also from the centre F a line FC is drawn to the point of contact C. Wherefore FC is perpendicular to KL,[c] whence each of the angles at C is a right angle. For the same reason the angles at the points B, D, are right ones. And because FCK is a right angle, the square of FK is equal to the squares of FC, CK.[d] For the same reason the square of FK is equal to the squares of FB, BK; wherefore the squares of FC, CK, are equal to the squares of FB, BK; of which the square of FC is equal to the square of FB; therefore the remaining square of CK is equal to the remaining square of BK; therefore CK is equal to BK. And because FB is equal to FC, and FK common, the two FB, BK, are equal to the two CF, CK, and the base BK is equal to the base CK; therefore the angle BFK is equal to the angle KFC,[e] also the angle BKF is equal to the angle FKC; therefore the angle BFC is double of the angle KFC, also the angle BKC of FKC. For the same reason CFD is double of CFL, and CLD of CLF. And because the circumference BC is equal to CD, the angle BFC is also equal to CFD.[f] And BFC is double of KFC, also DFC is double of LFC. Therefore KFC is also equal to LFC, and the angle FCK is equal to FCL. Whence there are two triangles FKC, FLC, having two angles equal to two angles, each to each, and one side equal to one side, viz. FC common to them; therefore the remaining sides will also be equal to the remaining sides, and the remaining angle equal to the remaining angle,[g] whence the right line KC is equal to CL, also the angle FKC to FLC. And because KC is equal to CL, therefore KL is double of KC. In the same manner it may be demonstrated that HK is double of BK. And BK is equal to KC; therefore KH is equal to KL. Similarly it may be demonstrated that each of the sides HG, GM, ML, is equal to each of the sides HK, KL. Therefore the pentagon GHKLM is equilateral. It is also equiangular. For because the angle FKC is equal to FLC, and it has been demonstrated that HKL is double of FKC, also KLM is

[c] 18. 3.
[d] 47. 1.
[e] 8. 1.
[f] 27. 3.
[g] 26. 1.

double of FLC; HKL is therefore equal to KLM. In like manner it may be demonstrated that each of the angles KHG, HGM, GML, is equal to each of the angles HKL, KLM; therefore the five angles GHK, HKL, KLM, LMG, MGH, are equal to one another. Therefore the pentagon GHKLM is equiangular. But it has been shown to be equilateral, and is circumscribed about the circle ABCDE. Q. E. F.

PROPOSITION XIII.

Problem.

To inscribe a circle in a given equiangular and equilateral pentagon.

Let ABCDE be a given equilateral and equiangular pentagon; it is required to inscribe a circle in the pentagon ABCDE.

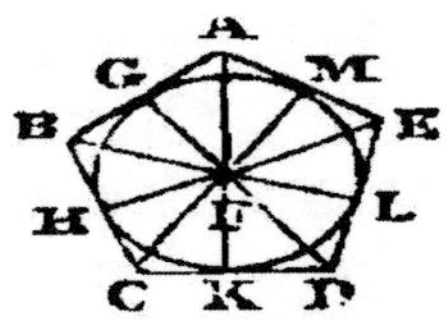

For bisect each of the angles BCD, CDE, by the right
a 9. 1. lines CF, DF;[a] and from the point F, in which the right lines CF, DF, meet one another, draw the right lines FB, FA, FE. And because BC is equal to CD, and CF common, the two BC, CF, are equal to the two DC, CF, and the angle BCF is equal to the angle
b 4. 1. DCF; therefore the base BF is equal to the base DF,[b] and the triangle BFC is equal to the triangle DFC, also the remaining angles will be equal to the remaining angles, which the equal sides subtend; therefore the angle CBF is equal to CDF. And because the angle CDE is double of CDF, and CDE is equal to ABC, also CDF to CBF, therefore CBA is double of CBF; whence the angle ABF is equal to FBC. Therefore the angle ABC is bisected by the right line BF. In like manner it may be demonstrated that each of the angles BAE, AED, is bisected by each of the right lines FA, FE. And draw from the point F to AB, BC, CD, DE, EA, the perpendicular right lines FG, FH, FK, FL, FM. And because the angle HCF is equal to KCF, and FHC is a right angle, and equal to the right angle FKC, so that there are two triangles FHC, FKC, having two angles equal to two angles, and one side equal to one side, viz. FC, which is common to both, subtending one of the equal angles; then the remaining sides shall be
c 26. 1. equal to the remaining sides;[c] therefore the perpen-

dicular FH is equal to the perpendicular FK. In like manner it may be demonstrated that each of the sides FL, FM, FG, is equal to each of the sides FH, FK; therefore the five right lines FG, FH, FK, FL, FM, are equal to one another. Wherefore with centre F and distance any one of them FG, FH, FK, FL, FM, the circle described will both pass through the remaining points, and touch the right lines AB, BC, CD, DE, EA; wherefore the angles at the points G, H, K, L, M, are right angles. For if it does not touch them, but cuts them, then a line drawn from the extremity at right angles to the diameter of the circle will fall within the circle, which has been shown to be absurd.[d] Therefore with centre F, and distance any one of the right lines FG, FH, FK, FL, FM, the circle described will not cut the right lines AB, BC, CD, DE, EA, whence it will touch them. Describe it as GHKLM, therefore in a given pentagon, &c. Q. E. F.

[d] 16. 3.

PROPOSITION XIV.

PROBLEM.

To circumscribe a circle about a given equilateral and equiangular pentagon.

Let ABCDE be the given equilateral and equiangular pentagon; it is required to circumscribe a circle about the given equilateral and equiangular pentagon ABCDE. Bisect each of the angles BCD, CDE, by the right lines CF, FD,[a] and from the point F, in which the right lines meet, draw to the points B, A, E, the right lines FB, FA, FE. In like manner, as has been before shown, that each of the angles CBA, BAE, AED, is bisected by each of the right lines FB, AF, EF. And because the angle BCD is equal to CDE, and FCD is half of BCD, also CDF is the half of CDE, and, therefore, FCD is equal to FDC; wherefore the side FC is equal to the side FD.[b] In like manner it may be demonstrated that each of the lines FB, FA, FE, is equal to each of the lines FC, FD; therefore the five right lines FA, FB, FC, FD, FE, are equal to one another. Therefore with centre F and distance any one of them FA, FB, FC, FD, FE, the circle described will

[a] 9. 1.

[b] 6. 1.

pass through the remaining points, and will be circumscribed, let it be circumscribed, and let it be ABCDE. Therefore a circle has been circumscribed, &c. Q. E. F.

PROPOSITION XV.

Problem.

To inscribe an equilateral and equiangular hexagon in a given circle.

Let ABCDEF be the given circle; it is required to inscribe an equilateral and equiangular hexagon in the circle ABCDEF.

Find G the centre of the circle, and draw AD the diameter of the circle ABCDEF, and with centre D, and
[a] 3 post. 1. distance DG, describe the circle EGCH,[a] also EG, GC, joined, produce to the points B, F, and join AB, BC, CD, DE, EF, FA, then is ABCDEF an equilateral and equiangular hexagon. For because the point G is the centre of the circle ABCDEF, GE is equal to GD. Again, because the point D is the centre of the circle EGCH, DE is equal to DG. But GE has been demonstrated to be equal to GD, therefore GE is equal to ED; whence EGD is an equilateral triangle, and, therefore, its three angles EGD, GDE, DEG, are equal to one another, because the angles at the
[b] 5. 1. base of an isosceles triangle are equal to one another.[b] And the three angles of a triangle are equal to two
[c] 32. 1. right angles;[c] therefore the angle EGD is a third part of two right angles. In like manner it may be demonstrated that DGC is a third part of two right angles. And because the right line CG standing upon EB makes the adjacent angles EGC, CGB, equal to two right an-
[d] 13. 1. gles,[d] and, therefore, the remaining angle CGE is a third part of two right angles; therefore the angles EGD, DGC, CGB, are equal to one another; and because the vertical angles BGA, AGF, FGE, are equal to EGD,
[e] 15. 1. DGC, CGB;[e] therefore the six angles EGD, DGC, CGB, BGA, AGF, FGE, are equal to one another. But equal
[f] 26. 3. angles stand upon equal circumferences;[f] therefore the six circumferences AB, BC, CD, DE, EF, FA, are equal to one another. And equal right lines subtend equal
[g] 29. 3. circumferences;[g] whence the six right lines are equal

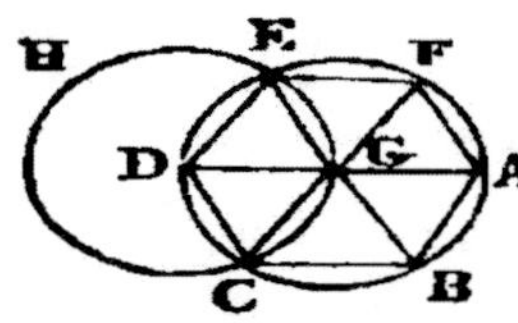

to one another; therefore the hexagon ABCDEF is equilateral. It is also equiangular; for because the circumference FA is equal to the circumference ED, add the circumference ABCD, which is common; therefore the whole FABCD is equal to the whole EDCBA, and the angle FED stands upon the circumference FABCD, also the angle AFE upon the circumference EDCBA. Therefore the angle AFE is equal to FED. In like manner it may be demonstrated that the remaining angles of the hexagon ABCDEF are each equal to each of the angles AFE, FED; therefore the hexagon ABCDEF is equiangular. But it has been shown to be equilateral, and is inscribed in the circle ABCDEF; wherefore an equilateral and equiangular hexagon has been inscribed in a given circle. Q. E. F.

Deduction.

To describe an equilateral and equiangular hexagon upon a given finite right line.

PROPOSITION XVI.

Problem.

To inscribe an equilateral and equiangular quindecagon in a given circle.

Let ABCD be the given circle; it is required to inscribe an equilateral and equiangular quindecagon in the circle ABCD.

Inscribe in the circle ABCD the side AC of an equilateral triangle, also AB the side of an equilateral pentagon. Therefore if such equal parts as the circumference ABCD contains fifteen, the circumference ABC, the third part of the whole, contains five; also AB, the fifth part of the whole, contains three; therefore the remainder BC contains two parts. Bisect BC in E,[a] therefore each of the circumferences BE, EC, will be the fifteenth of the whole ABCD. If, therefore, the right lines BE, EC, be drawn, and right lines equal to them be continually applied in the circle ABCD, there will be inscribed in it an equi- [a] 30. 3.

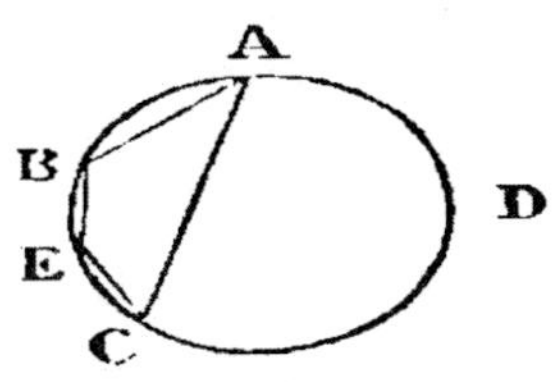

lateral and equiangular quindecagon.* Q. E. F. And in like manner as was done in the pentagon, if through the points of division right lines be drawn touching the circle, there will be circumscribed about the circle an equilateral and equiangular quindecagon; and, also, as in the pentagon, a circle may be inscribed in a given equilateral and equiangular quindecagon, and may be circumscribed about it.

---

* It was generally supposed that, besides the polygons here mentioned, no other could be inscribed by the scale and compasses only; until, at length, M. Gauss proved, in a work entitled *Disquisitiones Arithmeticæ*, published at Leipsig in 1801, and translated into French by M. Delisle, that a polygon of seventeen sides might be inscribed by the method in question, and, generally, any polygon, the number of whose sides is a prime number of the form $2^n + 1$.

EUCLID'S ELEMENTS.

BOOK V.

DEFINITIONS.

1. A magnitude is a part of a magnitude, a less of a greater, when the less measures the greater.
2. A multiple is a greater magnitude of a less, when the less measures the greater.
3. Ratio is a certain mutual habitude or relation of two magnitudes of the same kind, according to quantity.*

---

* Several mathematicians have found fault with this definition of Euclid, considering it obscure and difficult to be understood. Among these, the Rev. Dr. Abram Robertson, Professor of Astronomy at Oxford, printed a neat and valuable paper in 1789, for the use of his classes, being a demonstration of that definition, in seven propositions, the substance of which is as follows. He first premises this advertisement:

"As demonstrations depending upon proportionality pervade every branch of mathematical science, it is a matter of the highest importance to establish it upon clear and indisputable principles. Most mathematicians, both ancient and modern, have been of opinion that Euclid has fallen short of his usual perspicuity in this particular. Some have questioned the truth of the definition upon which he has founded it, and almost all who have admitted its truth and validity have objected to it, as a definition. The author of the following propositions ranks himself amongst objectors of the last mentioned description. He thinks that Euclid must have founded the definition in question upon the reasoning contained in the first six demonstrations here given, or upon a similar train of thinking, and in his opinion a definition ought to be as simple, or as free from a multiplicity of conditions, as the subject will admit."

He then lays down these four definitions:

"1. Ratio is the relation which one magnitude has to another of the same kind, with respect to quantity."

"2 If the first of four magnitudes be exactly as great when compared to the second, as the third is when compared to the fourth, the first is said to have to the second the same ratio that the third has to the fourth."

"3. If the first of four magnitudes be greater, when compared to the

4. Magnitudes are said to have a proportion to one another, which multiplied can exceed each other.

5. Magnitudes are said to be in the same ratio, the first to the second as the third to the fourth, when the equimultiples of the first and third compared with the equimultiples of the second

---

second, than the third is when compared to the fourth, the first is said to have to the second a greater ratio than the third has to the fourth."

"4. If the first of four magnitudes be less, when compared to the second, than the third is when compared to the fourth, the first is said to have to the second a less ratio than the third has to the fourth."

Dr. Robertson then delivers the propositions, which are the following:

"Prop. 1. If the first of four magnitudes have to the second the same ratio which the third has to the fourth; then, if the first be equal to the second, the third is equal to the fourth; if greater, greater; if less, less."

"Prop. 2. If the first of four magnitudes be to the second as the third to the fourth, and if any equimultiples whatever of the first and third be taken, and also any equimultiples of the second and fourth; the multiple of the first will be to the multiple of the second as the multiple of the third to the multiple of the fourth."

"Prop. 3. If the first of four magnitudes be to the second as the third to the fourth, and if any like aliquot parts whatever be taken of the first and third, and any like aliquot parts whatever of the second and fourth, the part of the first will be to the part of the second as the part of the third to the part of the fourth."

"Prop. 4. If the first of four magnitudes be to the second as the third to the fourth, and if any equimultiples whatever be taken of the first and third, and any whatever of the second and fourth; if the multiple of the first be equal to the multiple of the second, the multiple of the third will be equal to the multiple of the fourth; if greater, greater; if less, less."

"Prop. 5. If the first of four magnitudes be to the second as the third is to a magnitude less than the fourth, then it is possible to take certain equimultiples of the first and third, and certain equimultiples of the second and fourth, such, that the multiple of the first shall be greater than the multiple of the second; but the multiple of the third not greater than the multiple of the fourth."

"Prop. 6. If the first of four magnitudes be to the second as the third is to a magnitude greater than the fourth, then certain equimultiples can be taken of the first and third, and certain equimultiples of the second and fourth, such that the multiple of the first shall be less than the multiple of the second, but the multiple of the third not less than the multiple of the fourth."

"Prop. 7. If any equimultiples whatever be taken of the first and third of four magnitudes, and any equimultiples whatever of the second and fourth; and if when the multiple of the first is less than that of the second, the multiple of the third is also less than that of the fourth; or if when the multiple of the first is equal to that of the second, the multiple of the third is also equal to that of the fourth; or if when the multiple of the first is greater than that of the second, the multiple of the third is also greater than that of the fourth; then, the first of the four magnitudes is to the second as the third is to the fourth."

These propositions are demonstrated by strict mathematical reasoning; the paper has been considerably enlarged by its learned author, and recently published in the Edinburgh Encyclopedia.

and fourth, according to any multiplication whatsoever, either together exceed, or are together equal, or are together deficient to each other.*

6. Magnitudes having the same ratio are called proportionals.

7. But when of equimultiples, the multiple of the first exceeds the multiple of the second, but the multiple of the third does not exceed the multiple of the fourth, then the first is said to have a greater ratio to the second, than the third has to the fourth.†

8. Proportion is a similitude of ratios.

9. Proportion consists of three terms at least.

10. If three magnitudes be proportionals, the first is said to have to the third a duplicate ratio of that which it has to the second.

11. If four magnitudes be proportionals, the first is said to have to the fourth a triplicate ratio of that which it has to the second, and so forwards, always more by one, as long as the proportion continues.

12. Magnitudes are called homologous when the antecedents are to the antecedents, as the consequents to the consequents.

13. Alternate ratio is the assumption of the antecedent to the antecedent, and of the consequent to the consequent. ‡

14. Inverse ratio is an assumption of the consequent as the antecedent, and so compared with the antecedent to the consequent. §

---

* See Euclid's other definition of proportion in the Seventh Book.

† Such as the ratios 3 : 1 and 10 : 7, for if the first and third be multiplied by 2 and the second and fourth by 4, there will result 6 : 4; 20 : 28; where the first 6 is greater than the second 4, whilst the third 20 is less than the fourth 28.

‡ If $\left.\begin{matrix} A : B :: C : D \\ 3 : 4 :: 6 : 8 \end{matrix}\right\}$ then alternately $\left\{\begin{matrix} A : C :: B : D \\ 3 : 6 :: 4 : 8 \end{matrix}\right.$

In alternate proportion, it is necessary that the four magnitudes be of the same kind. For if a line A be to a line B, as a number C is to a number D, it does not follow that the line A will be to the number C as the line B to the number D, since no ratio between a line and number can be assigned.

§ If $\left.\begin{matrix} A : B :: C : D \\ 3 : 4 :: 6 : 8 \end{matrix}\right\}$ then inversely $\left\{\begin{matrix} B : A :: D : C \\ 4 : 3 :: 8 : 6 \end{matrix}\right.$

15. Composition of ratio is the assumption of the antecedent together with the consequent taken as one, to that consequent. *

16. Division of ratio is the assumption of the excess, by which the antecedent exceeds the consequent, to that consequent. †

17. Conversion of ratio is the assumption of the antecedent to the excess, by which the antecedent exceeds that consequent. ‡

18. Ratio of equality is when there are several magnitudes, and as many others, so that the first of the first magnitudes shall be to the last, as the first in the second magnitudes to the last. Or otherwise, the assumption of the extremes by subtracting the means. §

19. Ordinate proportion is, when it shall be as antecedent to a consequent, so is an antecedent to a consequent, and as the consequent is to any other, so is the consequent to any other. ||

20. Perturbate proportion is, when there are three magnitudes and others equal to them in number, it shall be as an antecedent in the first magnitudes is to a consequent, so is an antecedent in the second magnitudes to a consequent. And as a consequent in the first magnitudes to another, so is some other in the second magnitudes to an antecedent.**

---

* If $\left.\begin{array}{l} A : B :: C : D \\ 3 : 4 :: 6 : 8 \end{array}\right\}$ then by composition $\left\{\begin{array}{l} A+B : B :: C+D : D \\ 3+4 : 4 :: 6+8 : 8 \end{array}\right.$

† If $\left.\begin{array}{l} A : B :: C : D \\ 3 : 4 :: 6 : 8 \end{array}\right\}$ then by division $\left\{\begin{array}{l} A-B : B :: C-D : D \\ 4-3 : 3 :: 8-6 : 6 \end{array}\right.$

‡ If $\left.\begin{array}{l} A : B :: C : D \\ 3 : 4 :: 6 : 8 \end{array}\right\}$ then by conversion $\left\{\begin{array}{l} A : A-B :: C : C-D \\ 3 : 4-3 :: 6 : 8-6 \end{array}\right.$

§ If $\left.\begin{array}{l} A : B :: D : E \\ B : C :: E : F \end{array}\right\}$ then from equality $\left\{ A : C :: D : F \right.$

On the supposition that A, B, and C, are magnitudes in one order, and D, E, and F, in another.

|| If $\left.\begin{array}{l} A : B :: a : b \\ B : C :: b : c \\ C : D :: c : d \end{array}\right\}$ then if the ratios are taken equal in a direct order, and that the extremes are proportional, viz. A : D :: a : d, it is called ordinate proportion.

** As suppose the magnitude A is to the magnitude B as the magnitude C is to the magnitude D; and again, suppose the consequent B is to some other magnitude E as some other magnitude F is to the antecedent C; then is this proportion called perturbate. For further elucidation, consult Fenn's Euclid, page 167.

AXIOMS.

1. "Equimultiples of the same, or of equal magnitudes, are equal to one another."
2. "Those magnitudes of which the same, or equal magnitudes, are equimultiples, are equal to one another."
3. "A multiple of a greater magnitude is greater than the same multiple of the less."
4. "That magnitude of which a multiple is greater than the same multiple of another, is greater than that other magnitude."

PROPOSITION I.

Theorem.

If there be any number of magnitudes equimultiples of as many other magnitudes, each of each; whatsoever multiple one magnitude is of one, the same multiple shall all be of all.

Let AB, CD, be any number of magnitudes, equimultiples of as many other magnitudes E, F, each of each; whatsoever multiple AB is of E, the same multiple AB, CD, together, is of E and F together.

For because AB is an equimultiple of E, and CD of F; as many magnitudes as are in AB equal to E, so many will there be in CD equal to F. Divide AB into parts equal to E, which let be AG, GB; also CD into parts equal to F, namely, CH, HD. Therefore the multitude of parts CH, HD, will be equal to the multitude of them AG, GB. And because AG is equal to E, also CH equal to F; AG, CH, will be equal[a] to E and F. For the same reason GB is equal to E, and HD to F; therefore GB, HD, will be equal to E and F: whence as many magnitudes as are in AB equal to E, so many are there in AB, CD, equal to E, F. Wherefore what multiple AB is of E, the same multiple will AB, CD, be of E, F. Therefore, if there be any magnitudes, &c. Q. E. D.*

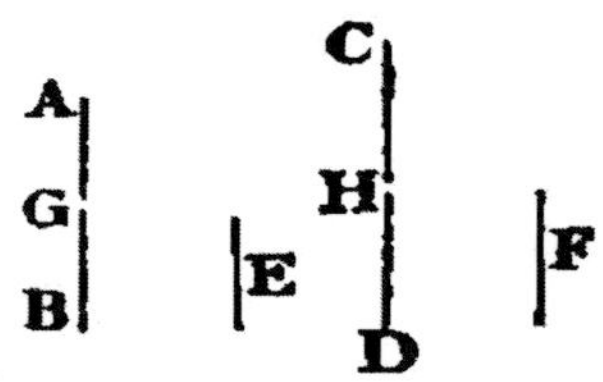

[a] Ax. 2. 1.

The same by Algebra.

Let there be any number of magnitudes am, an, equimultiples of as many others m, n; then shall am be the same multiple of m as $am + an$ is of $m + n$. For am is contained a times in m, and $am + an$ is also contained a times in $m + n$. Q. E. D.

---

* This is only a particular case of proposition 12.

PROPOSITION II.

THEOREM.

If the first magnitude be the same multiple of the second as the third is of the fourth, and the fifth be the same multiple of the second as the sixth is of the fourth; then the first and fifth taken together will be the same multiple of the second as the third and sixth are of the fourth.

For let the first magnitude AB be the same multiple of the second C, as the third DE is of the fourth F; and a fifth magnitude BG be the same multiple of the second C, as the sixth EH is of the fourth F: then is AG the first and fifth taken together the same multiple of the second C, as DH the third and sixth together is of the fourth F.

For because AB is the same multiple of C as DE is of F; as many magnitudes as are in AB equal to C, so many will there be in DE equal to F. And for the same reason as many magnitudes as are in BG equal to C, so many will there be in EH equal to F: therefore as many magnitudes as are in the whole AG equal to C, so many will there be in the whole DH equal to F.[a] Wherefore [a] 1. 5.

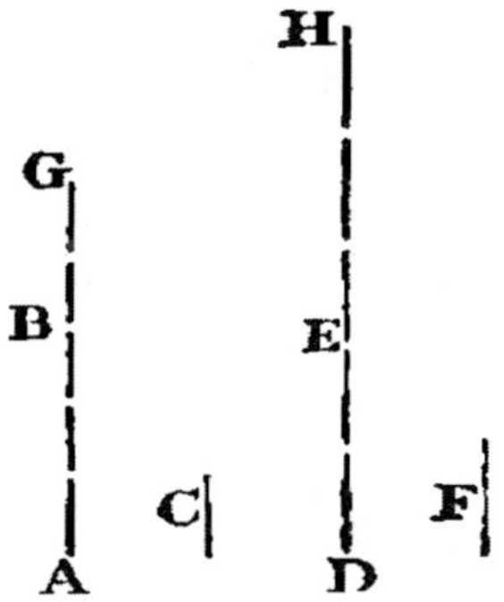

whatever multiple AG is of C, the same multiple is DH of F: therefore AG the first and fifth taken together shall be the same multiple of the second C, as DH the third and sixth, is of F the fourth. Wherefore if the first be the same multiple, &c. Q. E. D.

The same by Algebra.

Let $a\,m$ and $a\,n$ be equimultiples of the magnitudes m, n; also $b\,m$, $b\,n$, equimultiples of the same magnitudes m, n; then shall $a\,m + b\,m$ be the same multiple of m, as $a\,n + b\,n$ is of n. For m is contained $\overline{a + b}$ times in $a\,m + b\,m$; and n is contained $\overline{a + b}$ times in $a\,n + b\,n$. Q. E. D.

PROPOSITION III.

Theorem.

If the first magnitude be the same multiple of the second as the third is of the fourth, and let equimultiples of the first and third be taken; then, by equality will each of the assumed equimultiples be an equimultiple of each, the one of the second, and the other of the fourth.

For let the first A be the same multiple of the second B as the third C is of the fourth D, and take EF, GH, equimultiples of A, C: then is EF the same multiple of B as GH is of D.

For because EF is the same multiple of A as GH is of C; as many magnitudes as are in EF equal to A, so many will there be in GH equal C. Divide EF into magnitudes EK, KF, equal to A; also divide CH into magnitudes equal to C; viz. GL, LH: therefore the multitude of EK, KF, will be equal to the multitude of GL, LH. And because A is the same multiple of B as C is of D; but EK is equal A, and GL to C; EK will be the same multiple of B as GL is of D. For the same reason KF is the same multiple of B as LH is of D. Therefore because the first EK is the same multiple of the second B as the third GL is of the fourth D; and the fifth KF is the same multiple of the second B as the sixth LH of the fourth D: the magnitude EF, the first and fifth together, is the same multiple of the second B, as G, H, the third and sixth together

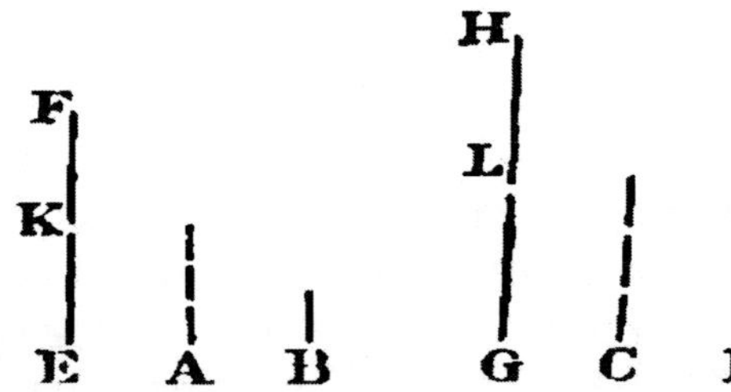

* 2. 5. is of the fourth D.[a] If, therefore, the first magnitude be the same multiple, &c. Q. E. D.

The same by Algebra.

Let the magnitude $b\,m$ be the same multiple of m, as $b\,n$ is of n; also let $a\,b\,m$, $a\,b\,n$, be equimultiples of the magnitudes $b\,m$, $b\,n$: then shall $a\,b\,m$ be the same multiple of m as $a\,b\,n$ is of n. For $a\,b\,m$ is contained $a\,b$ times in m, and likewise $a\,b\,n$ is contained $a\,b$ times in n. Q. E. D.

PROPOSITION IV.

THEOREM.

If the first magnitudes have the same ratio to the second as the third has to the fourth, then any equimultiples of the first and third will have the same ratio to any equimultiples of the second and fourth, according to any multiplication whatsoever, when compared with one another.

For let the first magnitude A have the same ratio to the second B, as the third C has to the fourth D, and take any how E, F, equimultiples of A, C, also others G, H, equimultiples of B, D: then as E is to C so is F to H.

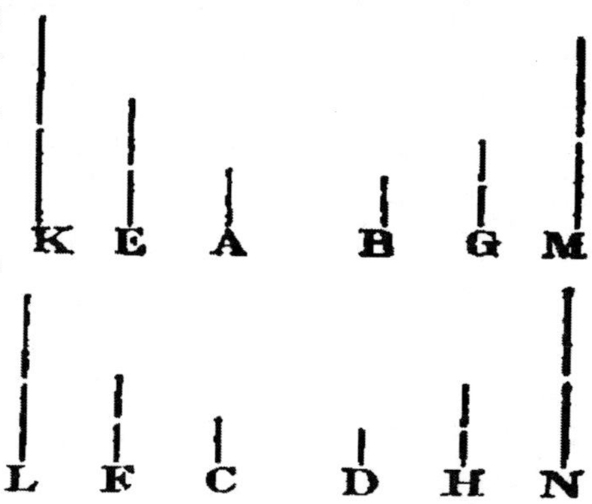

For take K, L, equimultiples of E, F, and M, N, equimultiples of G, H. Therefore because E is the same multiple of A as F is of C, and K, L, equimultiples of E, F, are taken: K will be the same multiple of A as L is of C. For the same reason M will be the same multiple of B as N is of D. And because A is to B as C is to D, and K, L, equimultiples of A, C, are taken, as also M, N, equimultiples of B, D: if K exceed M, L will exceed N; and if equal, equal; if less, less. And K, L, are equimultiples of E, F; also M, N, any other equimultiples of G, H; therefore as E is to G so will F be to H.[a] Wherefore, if the first have the same ratio, &c. Q. E. D. [a] 5 Def. 5.

The same by Algebra.

Let the magnitude m have the same ratio to n, as p has to q; and let $a\,m$ and $a\,p$ be any equimultiples of m and p, also $b\,n$ and $b\,q$ equimultiples of n and q: then will $a\,m$ have the same ratio to $b\,n$, as $a\,p$ has to $a\,q$; or $a\,m : b\,n :: a\,p : b\,q$, or $\frac{a\,m}{b\,n} = \frac{a\,p}{b\,q}$. For because $\frac{m}{n} = \frac{p}{q}$; multiply each side of the equation by $\frac{a}{b}$ and it will be $\frac{a\,m}{b\,n} = \frac{a\,p}{b\,q}$.* Q. E. D. * Ax. 1. 5.

Deduction.

If the first of four magnitudes have the same ratio to the second, which the third has to the fourth, then shall any equimultiples of the first and second have the same ratio which the third has to the fourth.

PROPOSITION V.

Theorem.*

If a magnitude be the same multiple of a magnitude, as a part taken away from the first is to a part taken away from the other, the remainder shall be the same multiple of the remainder as the whole is of the whole.

For let the magnitude AB be the same multiple of the magnitude CD, as AE, a part taken away from AB, is to CF, a part taken away from CD, then is the remainder EB the same multiple of the remainder FD as the whole AB is of the whole CD.

For whatsoever multiple AE is of CF, make EB the same multiple of CG.

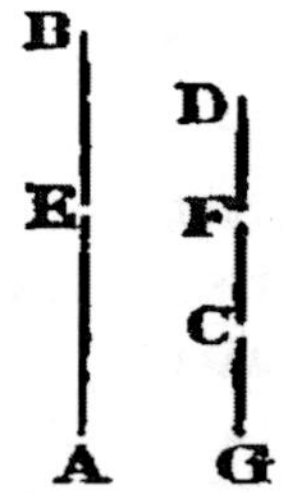

[a] 1. 5. And because AE[a] is the same multiple of CF as AB is of CD, and AE is put the same multiple of CF, as AB is of CD; AB is the same multiple of GF or CD: and consequently GF is equal to CD. Take away the common part CF: [b] 1 Ax. 5. therefore the remainder GC is equal[b] to the remainder DF. Whence, because AE is the same multiple of CF as EB is of GC, and GC is equal to DF, AE will be the same multiple of CF as EB is of FD. But AE is the same multiple of CF as AB of CD: therefore EB is the same multiple of FD as AB is of CD; whence the remainder EB is the same multiple of FD as the whole AB is of the whole CD. Q. E. D.

The same by Algebra.

Let the magnitude $a\,m$ be the same multiple of m as $a\,n$, a part taken from the first, is of n a part taken from the second; then shall the remainder, viz. $a\,m - a\,n$, be the same multiple of the remainder, viz. $m - n$, as the whole $a\,m$ is of the whole m. For $a\,m - a\,n$ is contained a times in $m - n$; also $a\,m$ is contained a times in m. Q. E. D.

---

* I have followed the Greek edition of Oxford, in making EB the same multiple of CG, as AE is of CF; which is likewise the case in Peyrard's edition.

PROPOSITION VI.

Theorem.

If two magnitudes be equimultiples of two others, and some parts taken away from them be equimultiples, then shall the remainders be either equal to these, or equimultiples of them.

For let there be two magnitudes, AB, CD, equimultiples of two others, E, F, and some parts, AG, CH, taken away from them, be equimultiples of them, E, F, then the remainders GB, HD, are either equal to E, F, or equimultiples of them.

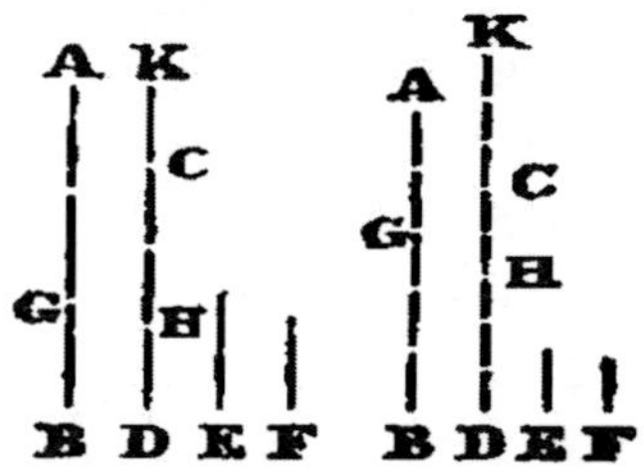

For, first, let GB be equal to E: then is HD equal to F. Make CK equal to F. And because AG is the same multiple of E as CH is of F, and GB is equal to E, CK also equal to F: AB[a] will be the same multiple of E as KH is of F. But AB is put the same multiple of E as CD is of F: therefore KH is the same multiple of F as CD is of F: whence because KH, CD, are each the same multiple of F, KH will be equal[b] to CD. Take away CH, which is common: therefore the remainder KC is equal to the remainder HD. But KC is equal to F: whence HD is equal to F. If, therefore, GB be equal to E, HD will also be equal to F.

[a] 2. 5.

[b] 1 Ax. 5.

In like manner, we demonstrate (as in fig. 2) if GB be a multiple of E, HD will be the same multiple of F. If, therefore, two magnitudes, &c. Q. E. D.

The same by Algebra.

Let the magnitudes am, an, be equimultiples of two others m, n, also bm, bn, some parts of the first magnitudes, be equimultiples of m, n, then shall the remainders, viz. $\overline{a-b}\,m$, $\overline{a-b}\,n$, be either equal to m, n, or equimultiples of them. For if $\overline{a-b}$ be equal to 1, it is manifest that $\overline{a-b}\,m$, $\overline{a-b}\,n$, are equal to m and n; but if not, then since $\overline{a-b}\,m$ is contained $\overline{a-b}$ times in m, and $\overline{a-b}\,n$ is contained $\overline{a-b}$ times in n, $\overline{a-b}\,m$ is the same multiple of m as $\overline{a-b}\,n$ is of n. Q. E. D.

PROPOSITION VII.

Theorem.

Equal magnitudes have the same ratio to the same magnitude, and the same magnitude has the same ratio to equal magnitudes.

Let A, B, be equal magnitudes, and C some other magnitude, then A, B, have the same ratio to C, and also C has the same ratio to either A or B.

For make D, E, equimultiples of A, B, and F any other multiple of C.

Then, because D is the same multiple of A as E is of B, and A is equal to B, D will also be equal to E. But F is any other multiple of C; therefore, if D exceed F, E will also exceed F; if equal, equal; and if less, less. And D, E, are equimultiples of A, B, but F some other multiple of C: therefore
a 5 Def. 5. A[a] will be to C as B is to C. Moreover, C has the same ratio to either A or B: for the same construction being made in like manner, we demonstrate D to be equal to E. If, therefore, F exceed D, it will also exceed E; and if equal, equal; if less, less. And F is any multiple of C; but D, E, any other equimul-
b 5 Def. 5. tiples of A, B: therefore as C[b] is to A, so will C be to B. Wherefore, equal magnitudes, &c. Q. E. D.

The same by Algebra.

Let m and n be equal magnitudes, and p some other magnitude; then $m : p :: n : p$; or, if $m : p :: n : p$, the magnitudes m and n are equal to one another. For take am, an, equimultiples of m and n; also cp, some multiple of p; then since m, n, are equal, am, an will also be equal; therefore, if am be greater, equal, or less,
* 5 Def. 5. than cp, an will* likewise be greater, equal, or less, than cp. But am, an, are equimultiples of the first term m, and of the third term n, also cp, cp, are equimultiples of the second and fourth terms p, p. Whence $m : p :: n : p$. Again, if $m : p :: n : p$, it follows that,
† 5 Def. 5. if am be greater, equal, or less, than cp, an will also † be greater, equal, or less, than cp. Therefore, $m = n$. Q. E. D.

PROPOSITION VIII.

THEOREM.

Of unequal magnitudes, the greater has a greater ratio to the same magnitude than the less has; and the same magnitude has a greater ratio to the less than it has to the greater magnitude.

Let AB, C, be unequal magnitudes, and let AB be greater than C; also let D be some other magnitude; then AB has a greater ratio to D than C has to D; also D has a greater ratio to C than it has to AB.

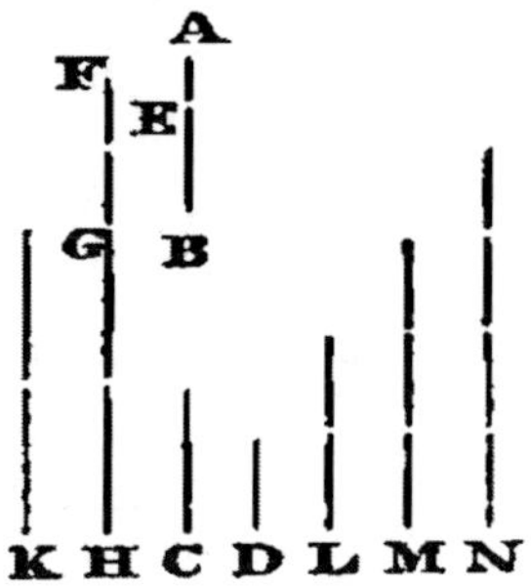

For, because AB is greater than C, make BE[a] equal to C; therefore, the less of AE, EB, being multiplied, will at length be greater than D. First, let AE be less than AB, and multiply AE until it becomes greater[b] than D; let FG be the multiple of AE, which is greater than D; then make GH the same multiple of EB, and K of C, as FG is of AE; and take L, double of D, M, triple of it, and so on, greater by one, until a multiple is taken greater than D, and in the first place greater than K. Let N be this magnitude, being quadruple of D, and in the first place greater than K. Therefore, because K is less than N, K will not be less than M. And since FG is the same multiple of AE as GH is of EB, FG[c] will also be the same multiple of AE as FH is of AB. But FG is the same multiple of AE as K is of C, and, consequently, FH, K, are equal multiples of AB, C. Again, because GH is the same multiple of EB as K is of C, and EB is equal to C, GH will also be equal to K. But K is not less than M; therefore GH is not less than M. But FG is greater[d] than D; whence the whole FH will be greater than D, M, together. But D, M, together are equal to N; wherefore FH exceeds N; but K does not exceed N: and FH, K, are equimultiples of AB, C, and N some other multiple of D. Therefore[e] AB has a greater ratio to D than C has to D.

[a] 3. 1.
[b] 4 Def. 5.
[c] 1. 5.
[d] By con.
[e] 7 Def. 5.

Moreover, D has a greater ratio to C, than D has to AB. For, the same construction being made, in like manner we demonstrate that N exceeds K, but does not exceed FH. And N is a multiple of D, also FH, K,

some other equimultiples of AB, C; therefore D has a
[f] 7 Def. 5. greater[f] ratio to C than D has to AB.

[g] 4 Def. 5. But if AB be greater than EB, therefore EB[g] the less may be multiplied until it be greater than D. Let it be multiplied, and let GH be that multiple of EB, greater than D. And what multiple GH is of EB; the same multiples make FG of AE, and K of C. For the same reason, when we before showed FH and K are equimultiples of AB, C. And in like manner, take N the same multiple of D, but in the first place greater than FG; therefore again FG is not less than M; but GH greater than D: therefore the whole FH exceeds D and M together, that is N. But K does not exceed N, because FG, which is greater than GH, that is, than K, does not exceed N. And in like manner, as before said, we finish the demonstration. Therefore of unequal magnitudes, the greater has, &c. Q. E. D.

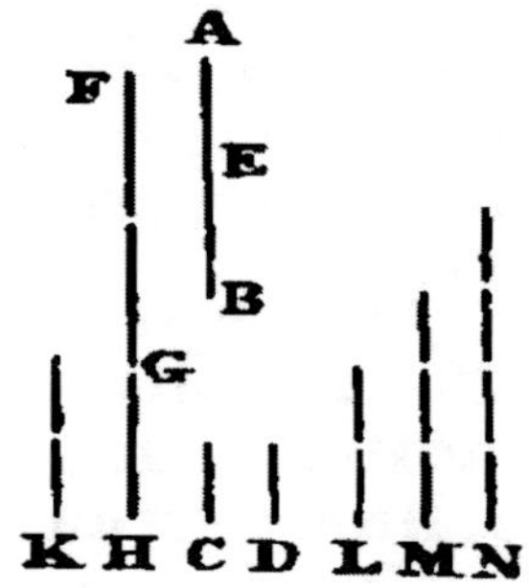

The same by Algebra.

Let m and n be two unequal magnitudes of which m is the greater, and let p be some other magnitude; then m has a greater ratio to p than n has; and p has a greater ratio to n than it has to m. Take a and b equimultiples of m and n, so that e, a multiple of p, may be greater than b, but less than a (which will easily happen if both a and b be taken greater than p). It is

* Def. 7. 5. evident that $\frac{m}{p}$* $> \frac{n}{p}$, and $\frac{p}{m} < \frac{p}{n}$. Q. E. D.

PROPOSITION IX.

Theorem.

Magnitudes which have the same ratio to the same magnitude, are equal to one another: and those to which the same magnitude has the same ratio are equal to one another.

For let each of the magnitudes A, B, have the same ratio to C: then is A equal to B.

For if they were not equal, A and B would not have

the same ratio[a] to C, but they have: therefore A is equal to B. [a] 8. 5.

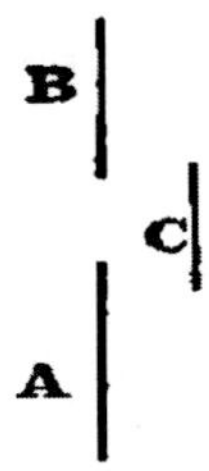

Again, let C have the same ratio to A and B, then A is equal to C.

For if it were not equal, C would not have the same ratio[a] to A, B, but they have: therefore A is equal to B. Therefore those magnitudes which have the same ratio to the same magnitude, &c. Q. E. D.

The same by Algebra.

Let m and n be two magnitudes, and let p be some other magnitude; then if $m : p :: n : p$; $m = n$. For $\frac{m}{p} = \frac{n}{p}$, $\therefore m = n$. Again, if $p : m :: p : n$; $m = n$. For $\frac{p}{m} = \frac{p}{n}$, or $n p = m p$, $\therefore n = m$. Q. E. D.

PROPOSITION X.

Theorem.

Of magnitudes having a ratio to the same magnitude, that which has the greater ratio, is the greater of the two; but to that which the same magnitude has the greater ratio, is the less of the two.

For let A have to C a greater ratio, than B has to C: then is A greater than B.

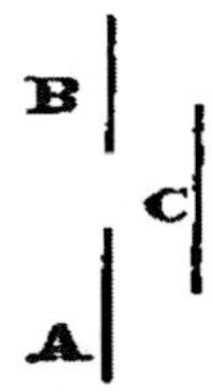

For if it be not greater, it is either equal or less. But it is not equal to B, for then each of the magnitudes A, B, would have the same proportion to C.[a] But they have not; therefore A is not equal to B, neither is A less than B, for then A would have a less proportion to C, than B has to C;[b] but it has not; therefore A is not less than B. And it has been shown not to be equal: wherefore A shall be greater than B. [a] 7. 5. [b] 8. 5.

Again, let C have a greater ratio to B than C has to A: then B is less than A.

For if it be not less, it is either equal or greater. B is not equal to A; for then C would have the same ratio to both A and B.[c] But it has not: wherefore A is not equal to B; neither is B greater than A; for then C would have a less ratio to B than it has to A.[d] But it has not: therefore B is not greater than A. And it has been shown not to be equal: wherefore B shall be less than A. Therefore of magnitudes having a ratio, &c. Q. E. D. [c] 7. 5. [d] 8. 5.

The same by Algebra.

Let m, n, be two magnitudes, and p some other magnitude, and let m have a greater ratio to p, than n has to p; then $m > n$, for $\frac{m}{p} > \frac{n}{p}$, $\therefore m > n$. Again, if p have a greater ratio to n than it has to m; $n < m$. For if not, it would be either equal or greater; but n does not equal m; for then $p : n :: p : m$; but it is not, neither is it greater; for then p would have a greater ratio to m than to n, but it has not: whence n is not greater than m, nor does $n = m$; $\therefore n < m$.

PROPOSITION XI.

Theorem.

Ratios which are the same to the same ratio are the same to one another.

For let A be to B as C is to D, and as C is to D so is E to F: then as A is to B so is E to F.

For take G, H, K, equimultiples of A, C, E; also any other L, M, N, equimultiples of B, D, F.

Therefore, because it is as A to B so is C to D, and G, H, are taken equimultiples of A, C, also L, M, any other equimultiples of B, D: if G exceed L, H will also
[a] 5 Def. 5. exceed M;[a] if equal, equal; and if less, less. Again, because it is as C is to D, so is E to F, and H, K, are taken equimultiples of C, E; also M, N, any other equimultiples of D, F: if H exceed M, K will also exceed N; if equal, equal; and if less, less.[a] But if H exceed M, G will also exceed L; if equal, equal; and if less, less. Wherefore, if G exceed L, K will also exceed N; and if equal, equal; and if less, less. And G, K, are equimultiples of A, E, also L, N, any other equimultiples of B, F: therefore as A is to B, so will E be to F.[a] Wherefore ratios which are the same to the same ratio, &c. Q. E. D.

The same by Algebra.

Let $a : b :: e : f$; and $c : d :: e : f$; then $a : b :: c : d$. Take g, h, i, equimultiples of a, c, e; and k, l, m, equi-
* Hyp. multiples of b, d, f. Because $a : b :: e : f$,* if $g <$,

$=, > k$,† then also $i <, =, > m$. And likewise be- † Def. 5.
cause $e : f :: c : d$,‡ if $h <, =, > l$,† then is i also ‡ Hyp.
$<, =, > m$:§ whence $a : b :: c : d$. Q. E. D.

PROPOSITION XII.

THEOREM.

If there be any number of magnitudes proportional; as one of the antecedents is to one of the consequents, so will all the antecedents be to all the consequents.

Let any number of magnitudes A, B, C, D, E, F, be proportional; that is, as A is to B, so is C to D, and E to F: as A : B, so are A, C, E, to B, D, F.

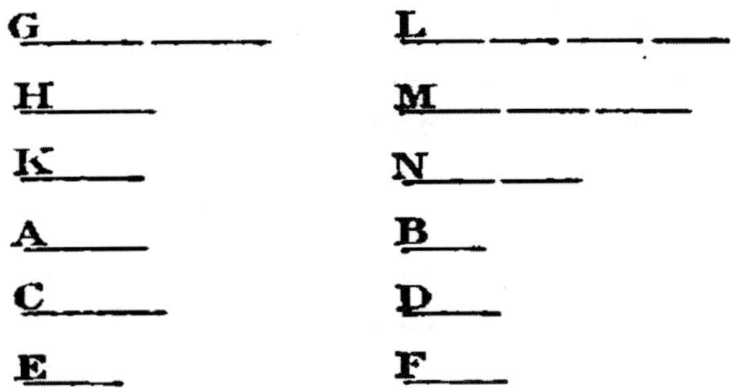

For take G, H, K, equimultiples of A, C, E; also L, M, N, any other equimultiples of B, D, F. Therefore, because as A is to B so is C to D and E to F; and G, H, K, have been taken equimultiples of A, C, E; also L, M, N, any other equimultiples of B, D, F: if G exceed L, H will also exceed M,[a] and [a] 5 Def. 5.
K exceed N; if equal, equal; and if less, less. Wherefore, if G exceed L; G, H, K, will also exceed L, M, N; if equal, equal; and if less, less: G, and G, H, K, are equimultiples of A, and A, C, E. Because[b] if there be [b] 1. 5.
any number of magnitudes equimultiples of as many magnitudes, each to each, whatsoever multiple any one of them is of its part, the same multiple shall all the first magnitudes be of all the others; for the same reason, L, and L, M, N, are equimultiples of B, and B, D, F: therefore it is[c] as A is to B, so are A, C, E, to [c] 5 Def. 5.
B, D, F. Wherefore, if there be any number, &c.§ Q. E. D.

The same by Algebra.

Let $s : t :: ms : mt :: ns : nt$, &c.; then will
$s : t :: s + ms + ns : t + mt + nt$, &c.

For $\frac{t + mt + nt}{s + ms + ns} = \frac{(1 + m + n)t}{(1 + m + n)s} = \frac{t}{s}$, the same ratio.

Deduction.

If any number of equal ratios be each greater than

§ This is the same as the twelfth proposition of the seventh book with regard to numbers.

a given ratio, the ratio of the sum of their antecedents, to the sum of their consequents, shall be greater than that given ratio.

PROPOSITION XIII.

Theorem.

If the first magnitude have the same ratio to the second, as the third has to the fourth; but the third have a greater ratio to the fourth, than the fifth has to the sixth: the first magnitude shall also have a greater ratio to the second than the fifth has to the sixth.

For let the first magnitude, A, have the same ratio to the second, B, as the third, C, to the fourth, D; but let the third, C, have a greater ratio to the fourth, D, than the fifth, E, to the sixth, F: then shall the first magnitude, A, also have a greater ratio to the second, B, than the fifth, E, to the sixth, F.

For because C has a greater ratio to D than E to F, there are some equimultiples of C, E, and some other

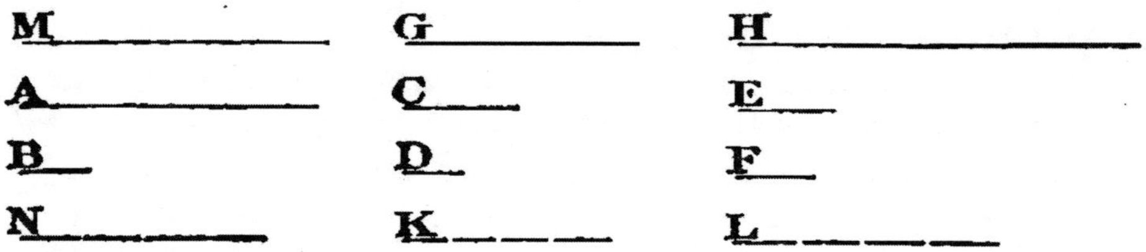

equimultiples of D, F, such, that the multiple of C is greater than the multiple of D, but the multiple of E
[a] 7 Def. 5. is not greater than the multiple of F.[a] And take G, H, equimultiples of C, E, also K, L, some other equimultiples of D, F, so that G shall exceed K, but H shall not exceed L; and whatever multiple G is of C, the same let M be of A; also whatever multiple K is of D, the same multiple let N be of B.

And because it is as A is to B so is C to D, and M, G, are taken equimultiples of A, C, also N, K, other equi-
[b] 5 Def. 5. multiples of B, D: if M exceed[b] N, G, will also exceed K; if equal, equal; and if less, less. But G exceeds K; therefore M also will exceed N. But H does not exceed L; and M, H, are equimultiples of A, E, and N, L, some other equimultiples of B, F: wherefore A shall have a greater ratio to B than E to F. If therefore the first have the same ratio, &c. Q. E. D.

The same by Algebra.

Let $a : b :: m\,a : m\,b$; also $\frac{m\,a}{m\,b} < \frac{c}{d}$; then will $\frac{a}{b} < \frac{c}{d}$. For $\frac{m\,a}{m\,b} < \frac{c}{d}$; or $\frac{a}{b} < \frac{c}{d}$. Q. E. D.

Deductions.

1. If the first of four magnitudes have a greater ratio to the second than the third has to the fourth; then shall the second have a less ratio to the first than the fourth to the third.

2. If any number of ratios be greater to the same ratio, then shall their sum be greater than that ratio multiplied by the number of the first ratios.

PROPOSITION XIV.

THEOREM.

If the first magnitude have the same ratio to the second, as the third has to the fourth, if the first be greater than the third, the second will also be greater than the fourth; if equal, equal; and if less, less.

For let A, the first magnitude, have to B, the second, the same ratio which C, the third, has to D, the fourth, and let A be greater than C; B will also be greater than D.

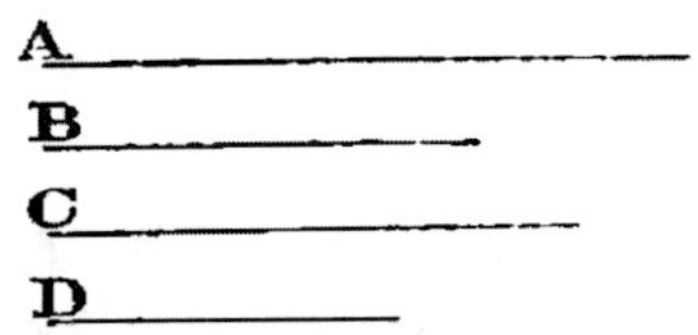

For because A is greater than C, and B some other magnitude; A shall have a greater ratio to B, than C to B.[a] But as A is to B, so is C to D; [a] 8. 5.
wherefore also C will have a greater ratio to D than C has to B.[b] But to that which the same magnitude has [b] 13. 5.
a greater ratio is the less:[c] whence D is less than B: [c] 10. 5.
and, consequently, B will be greater than D.

In like manner we demonstrate, that if A be equal to C, B will also be equal to D; and if A be less than C, B will also be less than D. If, therefore, the first magnitude have the same ratio, &c. Q. E. D.

The same by Algebra.

Let $a : b :: c : d$; or $\frac{a}{b} = \frac{c}{d}$; then if $a <, =,$ or $> c$, b shall also be $<, =,$ or $>$ than d. Let $a < c$,* * 8. 5.
then $\frac{a}{b} < \frac{c}{b}$, but $\frac{a}{b} = \frac{c}{d}$; whence,† $\frac{c}{d} < \frac{c}{b}$, therefore, † 13. 5.
$b < d$. In the same manner, if $a > c$,‡ then is $b > d$. ‡ 10. 5.
But if $a = c$, then $a : b :: c : b$. So that $b = d$. Q. E. D.

PROPOSITION XV.

Theorem.

Magnitudes, when compared to one another, have the same ratio as their equimultiples have to one another.

For let AB be the same multiple of C, as DE is of F, then as C is to F, so is AB to DE.

For because AB is the same multiple of C as DE is of F; as many magnitudes as are in AB equal to C, so many will there be in DE equal to F. Divide AB into magnitudes each equal to C, which let be AG, GH, HB; and DE into magnitudes each equal to F, viz. in DK, KL, LE. Therefore the multitude in AG, GH, HB, will be equal to the multitude of DK, KL, LE. And because AG, GH, HB, are equal to one another, also DK, KL, LE, are equal to one another, they will be as AG is to DK, so

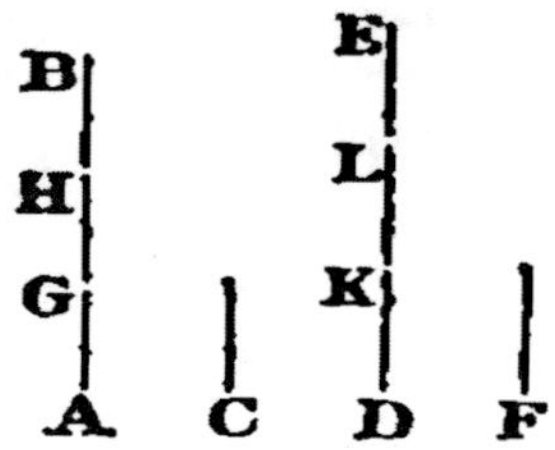

a 7. 5. is GH to KL, and HB to LE.[a] And it will be as one of the antecedents is to one of the consequents, so

b 12. 5. are all the antecedents to all the consequents:[b] therefore it is as AG is to DK so is AB to DE. But AG is equal to C, and DK to F: therefore as C is to F so will AB be to DE. Therefore magnitudes, when compared to one another, &c. Q. E. D.

The same by Algebra.

Let a and b be any two quantities, and $m\,a$, $m\,b$, any equimultiples of them, m being any number whatever: then will $a : b :: m\,a : m\,b$. For $\frac{m\,a}{m\,b} = \frac{a}{b}$. Q. E. D.

PROPOSITION XVI.

Theorem.

If four magnitudes of the same kind be proportional, they shall also be alternately proportional.

Let the four magnitudes A, B, C, D, be proportionals, viz. as A is to B so is C to D. They shall also be alternately proportional; viz. as A is to C so is B to D.

For take E, F, equimultiples of A, B; also G, H, any other equimultiples of C, D.

And because E is the same multiple of A as F is of B; and magnitudes when compared to one another have the same ratio as their equimultiples have to one another; it will be as A is to B so is E to F. But as A is to B so is C to D; therefore as C is to D so is E to

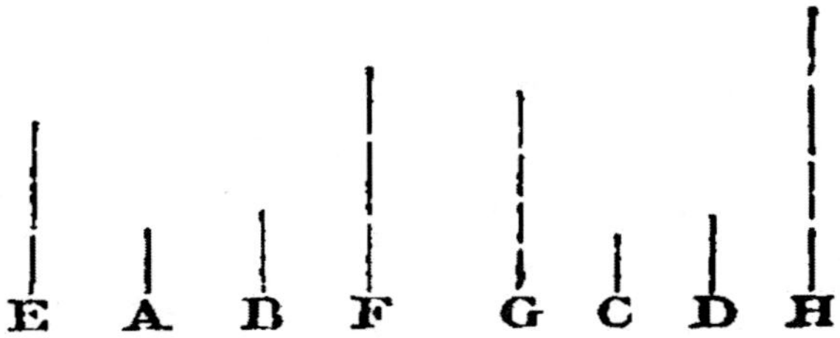

F.[a] Again, because G, H, are equimultiples of C, D; it will be as C is to D so is G to H. But as C is to D so is E to F: therefore as E is to F so is G to H.[a] But if four magnitudes be proportional, and the first be greater than the third; then the second will be greater than the fourth;[b] if equal, equal; and if less, less. If therefore E exceed G, F will also exceed H; if equal, equal; and if less, less. And E, F, are equimultiples of A, B, and G, H, some other equimultiples of C, D; therefore as A is to C so will B be to D.[c] If, therefore, four magnitudes be proportional, &c. Q. E. D.

[a] 11. 5.
[b] 14. 5.
[c] 5 Def. 5.

The same by Algebra.

Let $a : b :: c : d$; then will $a : c :: b : d$; or $\frac{a}{c} = \frac{b}{d}$.

For because $\frac{a}{b} = \frac{c}{d}$; multiply each side of the equation by $\frac{b}{c}$, and it will be $\frac{a\,b}{b\,c} = \frac{b\,c}{c\,d}$* or $\frac{a}{c} = \frac{b}{d}$. Q. E. D.

* 1 Ax. 5.

Deduction.

If the first of four magnitudes of the same kind have a greater ratio to the second than the third has to the fourth; the first shall also have a greater ratio to the third than the second has to the fourth.

PROPOSITION XVII.

THEOREM.

If magnitudes when compounded be proportional, they will also be proportional when divided.

Let the compounded magnitudes AB, BE, CD, DF, be proportional, and let it be as AB is to BE so is CD to DF: they are also proportional when divided; viz. as AE is to EB so is CF to FD.

For take GH, HK, LM, MN, equimultiples of AE, EB, CF, FD; also KX, NP, any other equimultiples of EB, FD.

And because GH is the same multiple of AE as HK is of EB; GH will also be the same multiple of AE as GK is of AB.[a] But GH is the same multiple of AE as LM is of CF: therefore GK is the same multiple of AB as LM is of CF. Again because LM is the same

[a] 1. 5.

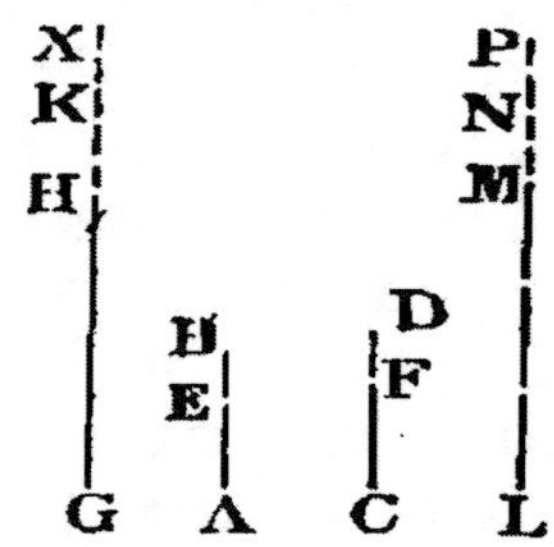

multiple of CF as MN is of FD; LM will also be the same multiple of CF as LN is of CD. But LM was the same multiple of CF as GK of AB: therefore GK is the same multiple of AB as LN is of CD; wherefore GK, LN, are equimultiples of AB, CD: again, because HK is the same multiple of EB as MN is of FD; and KX is the same multiple of EB as NP is of FD; also HX compounded is the same multiple of EB as MP compounded is of FD.[b] But since as AB is to BE as CD is to DF, and GK, LN, are taken equimultiples of AB, CD, also HX, MP, any other equimultiples of BE, FD: then if GK exceeds HX, LN will also exceed MP; if equal, equal; and if less, less.[c] Therefore let GK exceed HX, taking away HK, which is common, and GH will exceed KX. But if GK exceeds HX, and LN exceeds MP; whence LN exceeds MP, and taking away MN, which is common, LM will, likewise, exceed NP: wherefore if GH exceeds KX, LM will also exceed NP. In like manner we demonstrate, that if GH be equal to KX, LM is also equal to NP; and if less, less. But GH, LM, are equimultiples of AE, CF, and KX, NP, any other equimultiples of AE, CF: therefore as AE is to EB so will CF be to FD.[c] If, therefore, magnitudes when compounded, &c. Q. E. D.

[b] 2. 5.

[c] 5 Def. 5.

The same by Algebra.

Let the magnitudes when compounded, viz. a, b, c, d, be proportional, that is $a : b :: c : d$; they will also be proportional when divided; viz. $a - b : b :: c - d : d$. For $\frac{a}{b} = \frac{c}{d}$ subtract one from each side of the equation, and it will be $\frac{a}{b} - 1 = \frac{c}{d} - 1$,* or $\frac{a-b}{b} = \frac{c-d}{d}$. Q. E. D.

* 3 Ax. 1.

PROPOSITION XVIII.

Theorem.

If magnitudes when divided be proportional; they will also be proportional when compounded.

Let the divided magnitudes AE, EB, CF, FD, be proportionals, viz. as AE to EB so is CF to FD; they are also proportional when compounded, viz. as AB to BE, so is CD to FD.

For if it be not as AB to BE so is CD to FD; it will be as AB to BE so is CD to a magnitude either less or greater than FD. First let it be to a less, namely, to DG. And because it is as AB to BE so is CD to DG, these compounded magnitudes are proportional: therefore they will also be proportional when divided:[a] whence it is as AE to EB so is CG to GD. But it is as AE to EB so is CB to FD: wherefore as CG to GD so is CF to FD.[b] But CG, the first magnitude, is greater than CF, the third; therefore also the second magnitude, GD, will be greater than the fourth, FD.[c] And it is likewise less, which is impossible: whence it is not as AB to BE so is CD to a magnitude less than FD. In like manner we show that it is not to a greater. Therefore it is as AB to BE so is CD to FD. Wherefore if divided magnitudes, &c. Q. E. D.

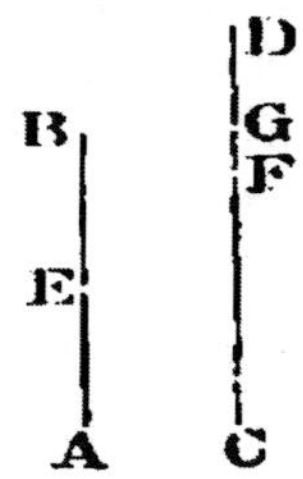

[a] 17. 5.
[b] 11. 5.
[c] 10. 5.

The same by Algebra.

Let the divided magnitudes a, b, c, d, be proportional, they are also proportional when compounded; viz. $a + b : b :: c + d : d$, or $\frac{a+b}{b} = \frac{c+d}{d}$. For since $\frac{a}{b} = \frac{c}{d}$,* add 1 to each side of the equation, and it will be $\frac{a}{b} + 1 = \frac{c}{d} + 1$;† or $\frac{a+b}{b} = \frac{c+d}{d}$ by reducing the quantities to improper fractions. Q. E. D.

* By hyp.
† 2 Ax. 1.

Deduction.

If the first of four magnitudes have a greater ratio to the second than the third has to the fourth, the first together with the second shall have a greater ratio to the second, than the third together with the fourth have to the fourth.

Dr. Simson has not only given us a very formidable demonstration of this proposition, but has also written nearly three pages in confuting the objection urged by Hieronymus Saccherius, and establishing his own: he says, "the present demonstration is none of Euclid's, nor is it legitimate; for it depends upon this hypothesis, that to any three magnitudes, two of which at least are of the same kind, there may be a fourth proportional." Dr. S. must surely have been aware that the theorem does not command us to find a fourth proportional to the three magnitudes in question; but only that such a proportional can possibly exist; and that it may, is evident if we conceive the third magnitude any multiple of the first, and also conceive some fourth magnitude the same multiple of the second; for then the four will be proportional by the fifteenth proposition; although it is certain that there are some magnitudes, which cannot be found to be exact multiples of other magnitudes; or, as they are called by mathematicians, incommensurables: yet the theorem itself will hold good even when applied to these incommensurables, as Dr. Gregory remarks in his edition of Hutton's Mathematics. It is true that even a conception of a thing founded upon an uncertain hypothesis ought never to be admitted, as it is not consonant with the rigid accuracy of geometrical reasoning, and is calculated to beget error: nevertheless, we find it frequently employed in modern systems of geometry; and Dr. Hutton, in the third proposition of his Geometry, says, "conceive the angle ACB to be bisected," without having shown us how it is bisected, or even the possibility of its being done.

PROPOSITION XIX.

Theorem.

If a whole magnitude be to a whole magnitude, as a part taken away from the first is to a part taken away from the other; then shall the remainder be to the remainder as the whole to the whole.

For let the whole magnitude AB be to the whole CD as AE, a part taken from the first magnitude, is to CF, a part taken from the other: then is the remainder EB to the remainder FD as the whole AB to the whole CD.

For because it is as AB to CD so is AE to CF; and,
[a] 16. 5. alternately, as BA is to AE so is DC to CF.[a] And because magnitudes when compounded are proportional, they shall also be proportional when
[b] 17. 5. divided;[b] therefore as BE is to

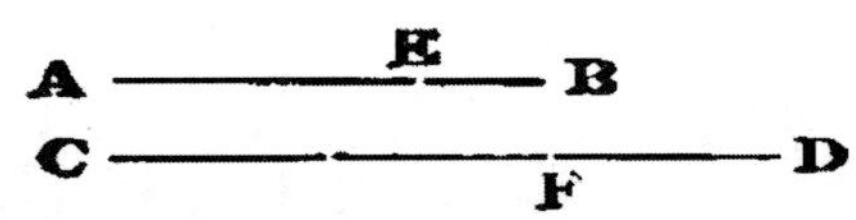

EA so is DF to FC; and, alternately, as BE is to DF so is EA to FC. But as AE to CF so is the whole AB to
[c] By hyp. the whole CD.[c] Therefore, the remainder EB will be to the remainder DF as the whole AB to the whole CD. If, therefore, a whole magnitude, &c. Q. E. D.

Corollary.

Hence it is manifest if compound magnitudes be proportional, they will also be proportional by conversion. For because it is as AB to CD so is AE to CF; and, alternately, as AB is to AE so is CD to CF; therefore magnitudes are proportional when compounded. But it has been shown as AB is to EB so is DC to FD, and it is by conversion. Q. E. D.

The same by Algebra.

Let a and b be two magnitudes, also c and d some other magnitudes, which are respective parts of a and b; then if $a : b :: c : d$; it shall also be $a - c : b - d :: a : b$. For as $a : b :: c : d$, alternately $a : c :: b : d$;* * 16. 5.
or $\frac{a}{c} = \frac{b}{d}$, subtract 1 from each side of the equation, and it will be $\frac{a}{c} - 1 = \frac{b}{d} - 1$;† or $\frac{a-c}{c} = \frac{b-d}{d}$; † 3 Ax. 1.
whence $a - c : c :: b - d : d$, and alternately $a - c : b - d :: c : d$; but $c : d :: a : b$; wherefore $a - c : b - d :: a : b$.‡ Q. E. D. ‡ 11. 5.

Deduction.

If any number of magnitudes be proportional, their differences will also be proportional.

PROPOSITION XX.

THEOREM.

If there be three magnitudes, and other three which taken two and two have the same ratio, then, if the first be greater than the third, the fourth also will be greater than the sixth; if equal, equal; and if less, less.

Let A, B, C, be three magnitudes, and D, E, F, other three, taken two and two, have the same ratio, viz. as A is to B so is D to E, also as B is to C so is E to F: then if A be greater than C, D also will be greater than F; if equal, equal; and if less, less.

For because A is greater than C, and B some other magnitude, and the greater magnitude has a greater ratio to the same than the less has;[a] therérefore A has a greater ratio to B than C has to B. a 8. 5.
But as A is to B so is D to E, and inversely as C is to B so is F to E; and D therefore has a greater ratio to E than F to E. But of those magnitudes having a ratio to the same magnitude, the greater has a greater ratio,[b] whence D is greater b 10. 5.
than F. In like manner we show, if A be equal to C, D will also be equal to F; and if less, less. If, therefore, there be three magnitudes, &c. Q. E. D.

The same by Algebra.

Let a, b, c, be three magnitudes, and d, e, f, other three, which taken two and two have the same ratio,

viz. $a : b :: d : e$, and $b : c :: e : f$; then if a be greater than c, d shall also be greater than f; if equal,
* Hyp. equal; and if less, less. If $a < c$, then because *$e : f :: b : c$, by inversion it will be $\therefore f : e :: c : b$. But
† 8. 5. $\frac{c}{b} > \frac{a}{b}$,† therefore $\frac{f}{e} > \frac{a}{b}$ or $\frac{d}{e}$, therefore $d < f$. In like manner we show, if $a > c$, also $d > f$, if $a = c$.
‡ 7. 5. Because $f : e :: c : b ::$ ‡$a : b :: d : e$§. Whence $d = f$.
§ 11. 5. Q. E. D.

PROPOSITION XXI.

THEOREM.

If there be three magnitudes, and others equal to them in number, which taken two and two have the same ratio, and let their proportion be perturbate; if the first magnitude be greater than the third, the fourth, also, will be greater than the sixth; if equal, equal; and if less, less.

Let A, B, C, be three magnitudes, and others D, E, F, equal to them in number, which taken two and two have the same ratio, and let their proportion be perturbate; viz. as A is to B so is E to F, and as B is to C so is D to E; then if A be greater than C, D will also be greater than F; if equal, equal; and if less, less.

A B C D E F

For because A is greater than C, and B some other magnitude; therefore A
a 8. 5. has a greater ratio to B than C has to B.[a] But as A is to B so is E to F; also inversely, as C is to B so is E to D; therefore also E has a greater ratio to F than E to D. But to that magni-
b 10. 5. tude, which the same has a greater ratio, is the less;[b] therefore F is less than D; whence D is greater than F. In like manner, we show if A be equal to C, D will also be equal to F; and if less, less. If, therefore, there be three magnitudes, &c. Q. E. D.

The same by Algebra.

Let a, b, c, be three magnitudes, and others d, e, f, equal to them in number, which taken two and two have the same ratio, viz. $a : b :: e : f$, and $b : c :: d : e$; if a be greater than c, d is also greater than f; if equal, equal; and if less, less.

1. If $a > c$, then because $d : e :: b : c$, therefore, inversely, $e : d :: c : b$, but $\frac{c}{b} < \frac{a}{b}$; whence $\frac{e}{d} < \frac{a}{b}$, that is, than $\frac{e}{f}$, therefore $d > f$.

2. In like manner, if $a < c$, then is $d < f$.

3. If $a = c$, then because $e : d :: c : b :: a : b :: e : f$, therefore is $d = f$. Q. E. D.

PROPOSITION XXII.

THEOREM.

If there be any number of magnitudes, and others equal to them in number, which taken two and two have the same ratio, they shall also be by equality in the same ratio, "that is, the first shall be to the last of the first magnitudes, as the first of the others is to the last."

Let there be any number of magnitudes A, B, C, and others equal to them in number, viz. D, E, F, which taken two and two have the same ratio, as A is to B so is D to E, and as B is to C so is E to F; then by equality, A shall be to C as D to F.

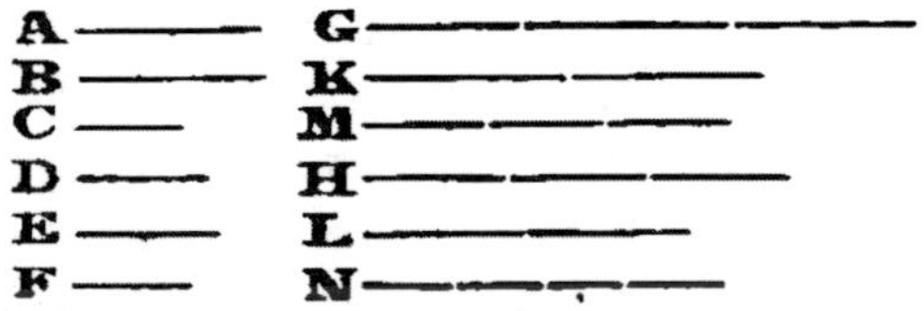

For take G, H, equimultiples of A, D, and K, L, any other equimultiples of B and E; moreover M, N, any other equimultiples of C, F. And because A is to B as D is to E; and G, H, are taken equimultiples of A, D, also K, L, any other equimultiples of B, E;[a] therefore as G is to K so is H to L. For the same reason as K is to M so is L to N. And because G, K, M, are three magnitudes, and others equal to them in number, viz. H, L, N, which taken two and two are in the same ratio; therefore by equality[b] if G exceeds M, H also exceeds N; if equal, equal; and if less, less. And G, H, are equimultiples of A, D, also M, N, any other equimultiples of C, F; therefore it is as A is to C so is D to F.[c] If, therefore, there be any number, &c. Q. E. D.

[a] 4. 5.
[b] 20. 5.
[c] 5 Def. 5.

The same by Algebra.

Let $a, b, c,$ be any magnitudes, and others $d, e, f,$ equal to them in number, which taken two and two have the same ratio; viz. $a : b :: d : e$, and $b : c :: e : f$; then by equality $a : c :: d : f$, or $\frac{a}{c} = \frac{d}{f}$. For because $\frac{a}{b} = \frac{d}{e}$, and $\frac{b}{c} = \frac{e}{f}$; multiply these two equations together, and it will be $\frac{a}{b} \times \frac{b}{c} = \frac{d}{e} \times \frac{e}{f}$,* or $\frac{ab}{bc} = \frac{de}{ef}$, or $\frac{a}{c} = \frac{d}{f}$. Q. E. D.

* 1 Ax. 5.

PROPOSITION XXIII.

Theorem.

*If there be three magnitudes, and others equal to them in number, which taken two and two have the same ratio, and their proportion be perturbate, they will also, by equality, be in the same ratio, "the first shall have the same ratio to the last of the first magnitudes, as the first of the others has to the last."**

Let A, B, C, be three magnitudes, and others D, E, F, equal to them in number, which taken two and two have the same ratio, and let their proportion be perturbate, viz. as A is to B so is E to F; also as B is to C so is D to E; then as A is to C so is D to F.

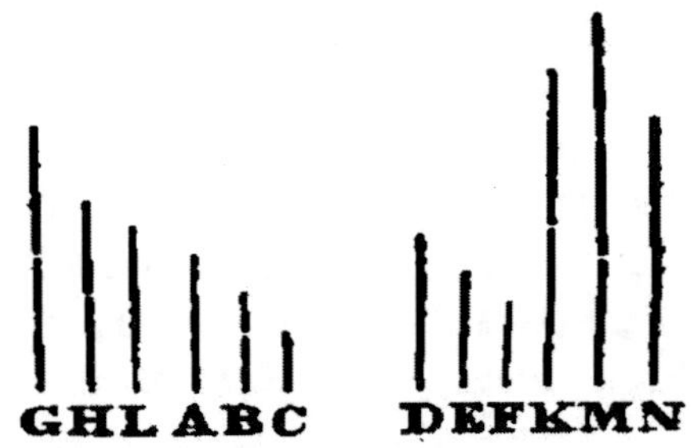

For take G, H, K, equimultiples of A, B, D, also L, M, N, any other equimultiples of C, E, F. And because G, H, are equimultiples of A, B, also that magnitudes have the same ratio which their equimultiples
a 15. 5. have;[a] therefore as A is to B so is G to H. By the same reason E is to F as M is to N; but it is as A is to B so is E to F; there-
b 11. 5. fore also as G is to H so is M to N.[b] And because it is as B is to C so is D to E, and alternately, as B is to D so is C to E. Also, because H, K, are equimultiples of B, D; and magnitudes have the same ratio which their equimultiples have; therefore as B is to D so is H to K; but as B is to D so is C to E; whence also, as H is to K so is C to E. Again, because L, M, are equimultiples of C, E; therefore it is as C is to E so is L to M. But as C is to E so is H to K; whence, also, as H is to K
c 15. 5. so is L to M, and, alternately, as H is to L so is K to M.[c] But it has been shown as G is to H so is M to N; and because there are three magnitudes G, H, L, and others, K, M, N, equal to them in number, taken two and two, have the same ratio, and their proportion is perturbate,
d 21. 5. therefore, by equality,[d] if G exceeds L, K also exceeds N; if equal, equal; and if less, less. And G, K, are equimultiples of A, D, also L, N, of C, F; therefore it
e 5 Def. 5. is as A is to C so is D to F.[e] If, therefore, there be three magnitudes, &c. Q. E. D.

* Euclid has demonstrated this proposition with proposing three magnitudes only; but this, as also the two following, will hold good with any number of magnitudes whatever.

The same by Algebra.

Let a, b, c, be three magnitudes, and d, e, f, as many others, which taken two and two have the same ratio; viz. $a : b :: e : f$, and $b : c :: d : e$; then $a : c :: d : f$, or $\frac{a}{c} = \frac{d}{f}$. For because $\frac{a}{b} = \frac{e}{f}$ and $\frac{b}{c} = \frac{d}{e}$; multiply these two equations together, and it will be $\frac{a}{b} \times \frac{b}{c} = \frac{e}{f}$* $\times \frac{d}{e}$ or $\frac{ab}{bc} = \frac{ed}{fe}$ or $\frac{a}{c} = \frac{d}{f}$. Q. E. D. * 1 Ax. 5.

PROPOSITION XXIV.

THEOREM.

If the first magnitude have the same ratio to the second which the third has to the fourth; and the fifth, the same ratio to the second, which the sixth has to the fourth; then the first and fifth taken together shall have the same ratio to the second which the third and sixth together have to the fourth.

For let the first magnitude, AB, have the same ratio to the second, C, which the third, DE, has to the fourth, F; and let the fifth, BG, have the same ratio to the second, C, which the sixth, EH, has to the fourth, F; then shall AG, the first and fifth taken together, have the same ratio to the second, C, which DH, the third and sixth together, have to the fourth, F.

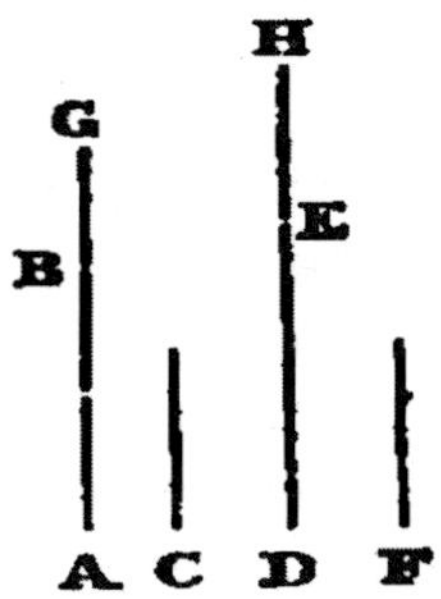

For because it is as BG to C so is EH to F; by inversion, therefore, as C is to BG so is F to EH. And because it is as AB to C so is DE to F, but as C to BG so is F to EH; therefore, by equality, it is as AB to BG so is DE to EH.[a] And because magnitudes divided are proportional, they shall also be proportional when compounded; therefore as AG is to BG so is DH to HE. But it is as BG to C so is EH to F; therefore, by equality, it is as AG to C so is DH to F.[b] If, therefore, the first magnitude, &c. Q. E. D. a 22. 5. b 18. 5.

The same by Algebra.

Let a the first magnitude have the same ratio to b the second, as c the third has to d the fourth; and let e the fifth have the same ratio to b the second, as f the sixth to d the fourth; or $a : b :: c : d$, and $e : b :: f : d$; then $a + e : b :: c + f : d$, or $\frac{a+e}{b} = \frac{c+f}{d}$.

For, because $\frac{a}{b} = \frac{c}{d}$ and $\frac{e}{b} = \frac{f}{d}$; add the two equations
* 2 Ax. 1. together and it will be $\frac{a}{b} + \frac{e}{b} = \frac{c}{d} + \frac{f}{d}$,* or $\frac{a+e}{b} = \frac{c+f}{d}$. Q. E. D.

PROPOSITION XXV.

THEOREM.

If four magnitudes be proportional, the greatest and least together are greater than the remaining two together.

Let the four magnitudes AB, CD, E, F, be proportional; viz. as AB is to CD so is E to F; let AB be the greatest, and consequently, F the least, then AB and F together are greater than CD and E together.

For make AG equal to E, also CH equal to F. Therefore because it is as AB is to CD so is E to F, but AG is equal to E and CH to F; therefore it is as AB is to CD so is AG to CH. And because it is as the whole magnitude AB is to the whole CD so is AG to CH, also
* 19. 5. the remainder GB[a] shall be to the remainder HD as the whole AB is to the whole CD. But AB is greater than CD; therefore GB is also greater than HD. And because AG is equal to E, also CH to F; therefore AG and F together are equal to CH and E together. And because, if equals be added to unequals, the wholes are unequal; if, therefore, GB, HD, being unequal, and CB being the greater; to GB add AG, F; also to HD add CH, E, therefore AB and F together will be greater than GD, E. If, therefore, four magnitudes, &c. Q. E. D.*

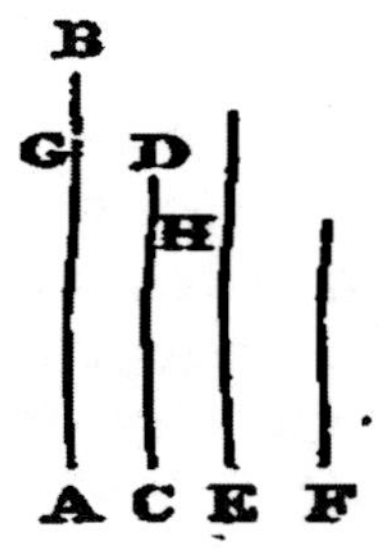

Deduction.

If the three magnitudes be proportional the two extremes shall be greater than double of the mean.

* From this it is manifest if the first term of the proportion be a maximum, the last will be a minimum.

EUCLID'S ELEMENTS.

BOOK VI.

DEFINITIONS.

1. Similar rectilineal figures are those which have their angles equal each to each, and the sides about the equal angles proportional.

2. Reciprocal figures are such as have their sides about two of their angles proportional in such a manner that a side of the first figure is to a side of the other, as the remaining side of this other is to the remaining side of the first.

3. A right line is said to be cut in extreme and mean ratio when the whole is to the greater segment as the greater segment is to the less.

4. The altitude of any figure is the perpendicular drawn from the vertex to the base.

PROPOSITION I.

THEOREM.

Triangles, and parallelograms, which have the same altitude, are to one another as their bases.

Let there be the triangles ABC, ACD, also the parallelograms EC, CF, which have the same altitude, viz. the perpendicular drawn from the point A to BD. Then as the base BC is to the base CD, so is the triangle ABC to the triangle ACD, and the parallelogram EC to the parallelogram CF.

Produce BD both ways to the points H, L, and take BG, GH, any number of times equal to the base BC; also DK, KL, any number of times equal to the base CD; and join AG, AH, AK, AL. Therefore because CB, BG, GH, are equal to one another, the triangles AHG, AGB, ABC, will be equal to one another.[a] Therefore the base HC

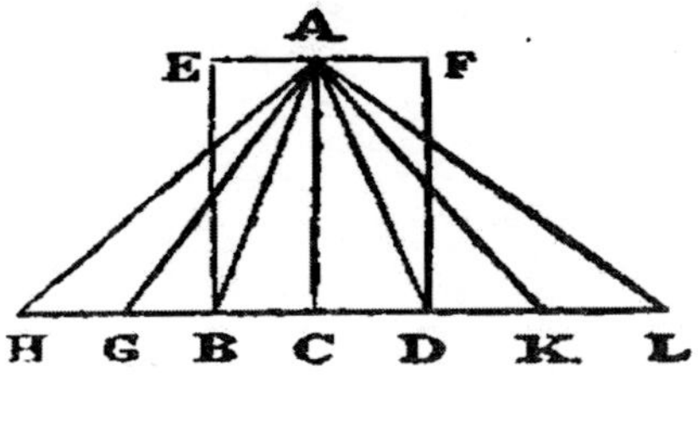

[a] 38. 1.

is the same multiple of the base BC, as the triangle AHC is of the triangle ABC. For the same reason the base LC is the same multiple of the base CD as the triangle ALC is of the triangle ACD; and if the base HC be equal to the base CL, the triangle AHC is equal to the triangle ALC: if the base HC be greater than the base CL, the triangle AHC is also greater than the triangle ALC; and if less, less. Therefore there are four magnitudes; viz. the bases BC, CD, and the two triangles ABC, ACD, such that if equimultiples of the base BC and the triangle ABC be taken, viz. the base HC and the triangle AHC; also of the base CD and the triangle ACD any other equimultiples, viz. the base CD and the triangle ALC. And it has been shown that if the base HC be greater than the base CL, the triangle AHC will be greater than the triangle ALC, if equal, equal, and if less, less; therefore as the base BC is to the base CD so
[b] 5 Def. 5. is the triangle ABC to the triangle ACD.[b]

And because the parallelogram EC is double of the
[c] 41. 1. triangle ABC[c] also the parallelogram FC is double of the triangle ACD; and magnitudes have the same ratio
[d] 15. 5. which their equimultiples have;[d] therefore as the triangle ABC is to the triangle ACD so is the parallelogram EC to the parallelogram FC. Hence because it has been shown, that as the base BC is to the base CD so is the triangle ABC to the triangle ACD; but as the triangle ABC is to the triangle ACD so is the parallelogram BC to the parallelogram FC; and, therefore, as the base BC is to the base CD so is the parallelogram EC to the
[e] 11. 5. parallelogram FC.[e] Therefore, triangles, &c. Q. E. D.

Deductions.

1. Triangles and parallelograms having equal bases, are to each other as their altitudes.

2. If four right lines be proportional, their squares shall also be proportional.

PROPOSITION II.

THEOREM.

If a right line be drawn parallel to one of the sides of a triangle, it shall cut the sides of the triangle proportionally; and if the sides of a triangle be cut proportionally, the right line joining the points of section shall be parallel to the remaining side of the triangle.

For draw DE parallel to one of the sides of the triangle ABC, viz. to BC; then as BD is to DA so is CE to EA.

Join BE, CD.

Then the triangle BDE is equal to the triangle CDE,[a] for they are upon the same base DE and between the same parallels DE, BC. But ADE is any other triangle; also equal magnitudes have the same ratio to the same magnitude;[b] therefore as the triangle BDE is to the triangle ADE so is the triangle CDE to the triangle ADE. But as the triangle BDE is to the triangle ADE so is BD to DA.[c] For they having the same altitude, viz. the perpendicular drawn from the point E to AB are to one another as their bases. For the same reason as the triangle CDE is to the triangle ADE so is CE to EA; and hence as BD is to DA so is CE to EA.[d] But also if the sides AB, AC, of the triangle ABC be cut proportionally in the points D, E, as BD is to DA so is CE to EA, and join DE; then is DE parallel to BC.

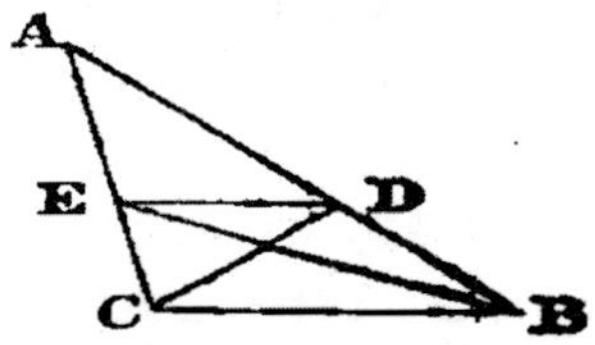

[a] 37. 1. [b] 7. 5. [c] 1. 6. [d] 11. 5.

For the same construction being made, because it is as BD is to DA so is CE to EA; but as BD is to DA so is the triangle BDE to the triangle ADE,[e] but as CE is to EA so is the triangle CDE to the triangle ADE; and therefore as the triangle BDE is to the triangle ADE so is the triangle CDB to the triangle ADE. Hence each of the triangles BDE, CDE, has the same ratio to the triangle ADE. Therefore the triangle BDE is equal to the triangle CDE,[f] and they are upon the same base DE. But equal triangles constituted upon the same base are also between the same parallels.[g] Therefore DE is parallel to BC. If, therefore, triangles, &c. Q. E. D.*

[e] 1. 6. [f] 9. 5. [g] 39. 1.

* From this and the 18th proposition of the fifth book, it may be demonstrated that the sides of the triangles AED, ACB, containing the angle CAB are proportional, viz. as AE is to AD so is AC to AB.

PROPOSITION III.

Theorem.

If the angle of a triangle be bisected, and the right line cutting the angle cuts the base also; the segments of the base shall have the same ratio which the remaining sides of the triangle have; and if the segments of the base have the same ratio which the remaining sides of the triangle have, the right line drawn from the vertex to the point of section bisects the vertical angle of the triangle.

Let ABC be a triangle and bisect the angle BAC by the right line AD; then as BD is to DC so is BA to AC.

[a] 31. 1. For through C draw CE parallel[a] to DA and BA produced will meet with it in E.

And because the right line AC falls upon the parallel right lines AD, EC, therefore the angle ACE is equal to [b] 29. 1. the angle CAD.[b] But CAD is supposed equal to BAD, therefore BAD is also equal to ACE. Again, because the right line BAE falls upon the parallels AD, EC, the exterior angle BAD is equal to the interior angle AEC. But it has been shown that ACE is equal to BAD, and hence the angle ACE is equal to AEC; wherefore, also, [c] 6. 1. the side AE is equal to the side AC.[c] And because AD is drawn parallel to one of the sides EC of the triangle BCE, therefore, proportionally, as BD is to DC so is BA to [d] 2. 6. AE.[d] And AE is equal to AC; therefore as BD is to DC [e] 7. 5. so is BA to AC.[e]

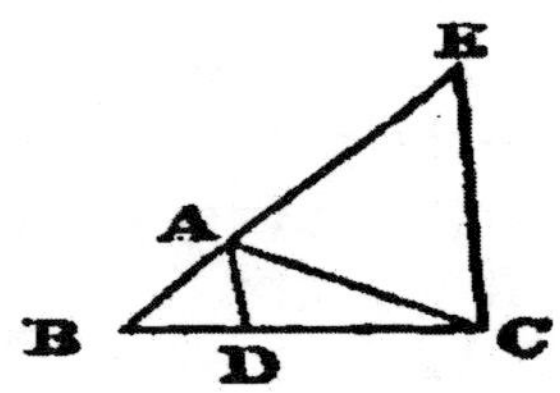

But let BD be to DC as BA is to AC, and join AD; then is the angle BAC bisected by the right line AD.

For the same construction being made, because as BD is to DC so is BA to AC, but also as BD is to DC so is BA to AE; for AD is drawn parallel to one of the sides EC of the triangle BCE; therefore, also, as BA is to AC so is BA to AE; hence AC is equal to AE; wherefore the angle AEC is equal to the angle ACE. But the angle AEC is equal to the exterior angle BAD, also the angle ACE to the alternate angle CAD, and therefore BAD is equal to CAD. Hence the angle BAC is bisected by the right line AD. If, therefore, the angle of a triangle, &c. Q. E. D.

PROPOSITION IV.

THEOREM.

The sides of the equiangular triangles about the equal angles are proportional, and the homologous sides subtend the equal angles.

Let ABC, DCE, be equiangular triangles, having the angle BAC equal to CDE, also ACB to DEC, and consequently ABC to DCE;[a] then the sides of the triangles ABC, DCE, about the equal angles, are proportional, and the homologous sides subtend the equal angles. [a] 32. 1.

For put BC in a direct line with CE, and because the angles ABC, ACB, are less than two right angles,[b] but ACB is equal to DEC, therefore ABC, DEC, are less than two right angles; hence BA, ED, produced, will meet.[c] Let them be produced, and let them meet in F. [b] 17. 1. [c] 12 Ax. 1.

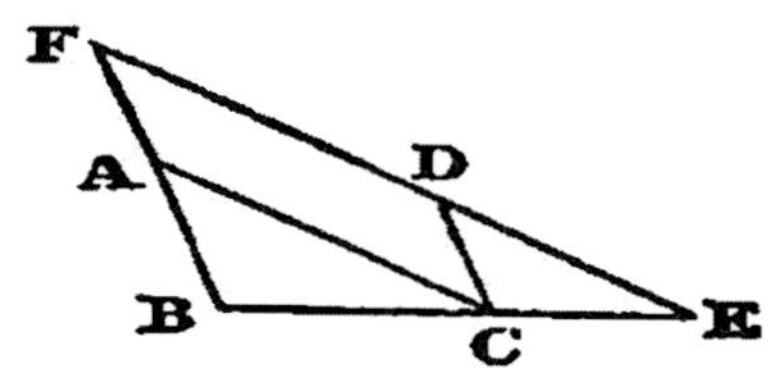

And because the angle DCE is equal to ABC, therefore BF is parallel to CD.[d] Again, because the angle ACB is equal to DEC, AC is parallel to FE; hence FACD is a parallelogram; therefore FA is equal to DC, also AC to FD.[e] And because AC is drawn parallel to one of the sides, FE, of the triangle FBE, therefore, as BA is to AF so is BC to CE.[f] But AF is equal to CD; hence as BA is to CD so is BC to CE, and alternately, as AB is to BC so is DC to CE. Again, because CD is parallel to BF, therefore as BC is to CE so is FD to DE. But FD is equal to AC; hence as BC is to CE so is AC to ED, and alternately, as BC is to CA so is CE to ED. And because it has been shown as AB is to BC so is DC to CE; also as BC is to CA so is CE to ED; therefore, by equality, as BA is to AC so is CD to DE. Therefore the sides of equiangular triangles about, &c. Q. E. D.* [d] 28. 1. [e] 34. 1. [f] 2. 6.

Deductions.

1. In isosceles triangles, which are equiangular, the perpendiculars drawn from the vertices to the bases are proportional to the sides of the triangle.

---

* Hence if in a triangle FBE, there be drawn AC, parallel to the side FE, the triangle ABC shall be similar to the whole FBE.

2. In equiangular triangles the radii of their inscribed circles have the same ratio as the sides of the triangle.

3. If a circle be touched in the same point, both externally and internally, by two other circles, and through the point of contact two right lines be drawn, the parts of them intercepted between the circumference of the given circle, and that of the circle which touches it internally, shall have to one another the same ratio as the parts which are the chords of the other circle.

4. The right lines, drawn from the bisections of the three sides of a triangle to the opposite angles, meet in the same point.

PROPOSITION V.

Theorem.

If two triangles have their sides proportional, the triangles shall be equiangular, and shall have those angles equal which subtend the homologous sides.

Let ABC, DEF, be two triangles, having their sides proportional, viz. as AB is to BC so is DE to EF, also, as BC is to CA so is EF to FD, and likewise as BA is to AC so is ED to DF; then the triangle ABC is equiangular to the triangle DEF, and have those angles equal which subtend the homologous sides, viz. ABC to DEF, BCA to EFD, and BAC to EDF.

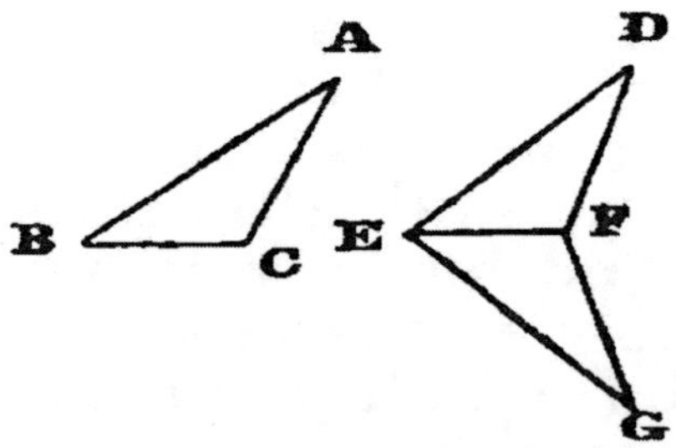

For to the right line EF, and at the points E, F, in it, make the
[a] 23. 1. angle FEG equal to the angle ABC,[a] also BCA to EFG, hence the remaining angle at A is equal to the
[b] 32. 1. remaining angle at G.[b] Therefore the triangle ABC is equiangular to EGF, and, consequently, the sides
[c] 4. 6. of the triangles ABC, EGF, are proportional,[c] and the homologous sides subtend the equal angles; therefore,
[d] 11. 5. as AB is to BC so is GE[d] to EF. But as AB is to BC so is DE to EF; and, therefore, as DE is to EF so is GE to EF; hence each of the sides DE, GE, have the same ratio to EF; therefore DE is equal to GE. For the same reason DF is equal to GF. And because DE is equal to EG, and EF is common, the two, DE, EF, are equal to the two, GE, EF, and the base FD is equal to the base FG; therefore the angle DEF is equal to the

angle GEF, and the remaining angles equal to the remaining angles, each to each, which subtend the equal sides; therefore the angle DFE is equal to GFE; also EDF to EGF. And because the angle FED is equal to the angle FEG, but FEG is equal to ABC, and, therefore, the angle ABC is equal to the angle DEF. For the same reason ACB is equal to DFE, and, likewise, A to D; hence the triangle ACB is equiangular to the triangle DEF. If, therefore, two triangles have their sides, &c. Q. E. D.

PROPOSITION VI.

THEOREM.

If two triangles have one angle of the one equal to one angle of the other, and the sides about the equal angles proportional, the triangles shall be equiangular, and have those angles equal which subtend the homologous sides.

Let ABC, DEF, be two triangles, having one angle BAC equal to one angle EDF, and the sides about the equal angles proportional, as BA is to AC so is ED to DF; then the triangle ABC is equiangular to the triangle DEF, and the angle ABC to DEF; also ACB to DFE.

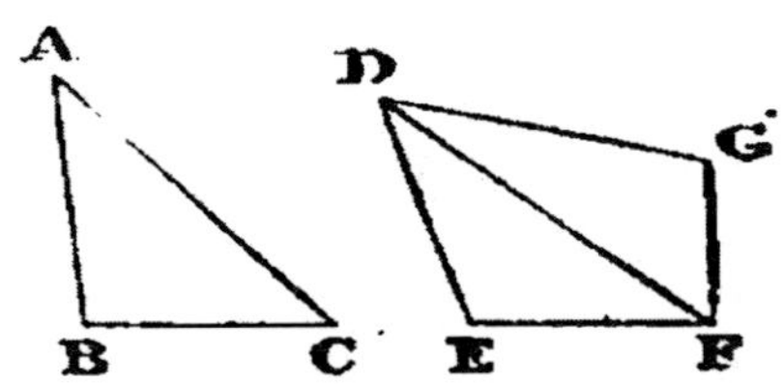

For to the right line DF, and at the points DF in it, make the angle FDG equal to each of them BAC, EDF; also DFG to ACB.[a] [a] 23. 1.

Therefore the remaining angle at B is equal to the remaining angle at G;[b] therefore the triangle ABC is equiangular to the triangle DGF; hence, proportionally, as BA to AC so is GD to DF.[c] But it is put, as BA to AC so is ED to DF, and, consequently, as ED is to DF so is GD to DF;[d] hence ED is equal to DG[e] and DF common: the two sides ED, DF, are equal to the two GD, DF, and the angle EDF is equal to GDF; hence the base EF is equal to the base FG, and the triangle DEF is equal to the triangle DGF: also the remaining angles to the remaining angles, those which subtend the equal sides; therefore the angle DFG is equal to the angle DFE; also DGF to DEF. But DGF is equal to ACB,

[b] 32. 1. [c] 4. 6. [d] 11. 5. [e] 9. 5.

and ACB, therefore, is equal to DFE. But BAC is equal to EDF, and hence the remaining angle at B is equal to the remaining angle at E; therefore the triangle ABC is equiangular to the triangle DEF. If, therefore, two triangles, &c. Q. E. D.*

PROPOSITION VII.

Theorem.

If two triangles have one angle of the one equal to one angle of the other, and the sides about two other angles proportional, then if each of the remaining angles be either less or not less than a right angle, the triangles shall be equiangular, and shall have those angles equal about which the sides are proportional.

Let ABC, DEF, be two triangles, having one angle equal to one angle, viz. BAC to EDF, and about two other angles, ABC, DEF, the sides proportional, viz. as AB to BC so is DE to EF; and first let each of the remaining angles at C, F, be less than a right angle, then the triangle ABC is equiangular to the triangle DEF, and the angle ABC equal to DEF; also the angle at C equal to the remaining angle at F.

For if the angle ABC be unequal to DEF, one of them is the greater; let ABC be the greater, and to the right line AB, and at the point in it B;
a 23. 11. make ABG equal to the angle DEF.[a] And because the angle A is equal to D, also the angle ABG to DEF, hence the remaining angle AGB is
b 32. 1. equal to the remaining angle DFE;[b] therefore the triangle ABG is equi-

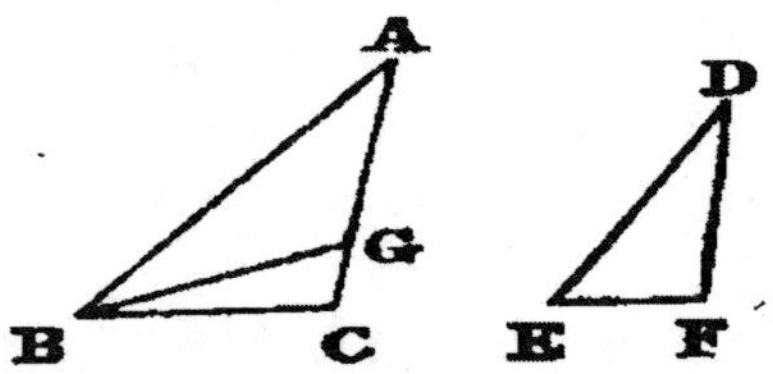

c 4. 6. angular to the triangle DEF, therefore as AB[c] is to BG so is DE to EF. But it is as DE to EF so is AB to
d 11. 5. BC; hence also as AB is to BC so is AB to BG,[d] there-fore AB has the same ratio to each of them BC, BG,
e 9. 5. consequently BC is equal to BG,[e] wherefore also the angle at C is equal to the angle BGC. But C is put less than a right angle, therefore BGC is less than a

---

* From this and the preceding proposition it evidently appears that the equality of angles in triangles is a consequence of the proportionality among the sides, so that one of these conditions being known is sufficient to determine whether the triangles are similar or not.

right angle, hence the adjacent angle AGB is greater than a right angle. And it has been shown to be equal to F, therefore F is greater than a right angle, but it is put less than a right angle, which is absurd; therefore the angle ABC is not unequal to DEF, and consequently it is equal. But the angle A is equal to D, and hence the remaining angle C is equal to the remaining angle F, therefore the triangle ABC is equiangular to the triangle DEF.

But again, let each of them C, F, be not less than a right angle; then again the triangle ABC is equiangular to DEF.

For the same construction being made, in like manner we demonstrate BC to be equal to BG, wherefore also the angle C is equal to BGC. But the angle at C is not less than a right angle, neither then is BGC less than a right angle, which is impossible; therefore again the angle ABC is not unequal to the angle DEF, that is, it is equal. But the angle at A is equal to the angle at D, therefore the remaining angle at C is equal to the remaining angle at F; hence the triangle ABC is equiangular to the triangle DEF. If, therefore, two triangles, &c. Q. E. D.

PROPOSITION VIII.

THEOREM.

If in a right angled triangle a perpendicular be drawn from the right angle to the base, the triangles made by the perpendicular are similar to the whole and to one another.

Let ABC be a right angled triangle, having the right angle BAC, and from A draw AD perpendicular to BC; then each of them ABD, ADC, is similar to the whole and to one another.

For because the angle BAC is equal to ABD, for each of them is a right one, and the angle at B is common to the two triangles ABC and ABD; hence the remaining angle ACB is equal to the remaining angle BAD,[a] [a] 32. 1.
and the triangle ABC is equiangular to the triangle ABD. Therefore as BC subtending the right angle of the triangle ABC is to BA, subtending the right angle of the triangle ABD, so is AB subtending the angle at C

of the triangle ABC to BD, subtending the angle equal to that at C, viz. BAD of the triangle ABD; and also AC is to AD subtending the angle at B, which is common to the two triangles, therefore, also the triangle ABC is equiangular to the triangle ABD, and has the sides about the equal angles proportional,[b] therefore the triangle ABC is similar to the triangle ABD. In like manner we show that the triangle ABC is similar to the triangle ADC, therefore each of the triangles ABD, ADC, is similar to the whole triangle ABC.

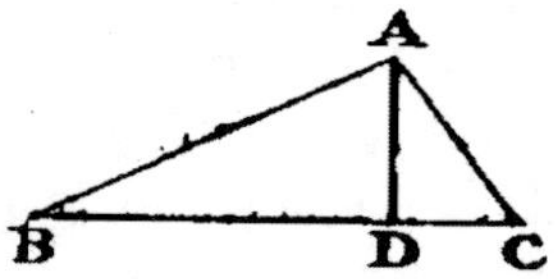

[b] 4. 6.

Again, ABD, ADC, are similar to one another.

For because the right angle BDA is equal to the right angle ADC, but also BAD has been shown to be equal to that at C, and therefore the remaining angle at B is equal to the remaining angle DAC; hence the triangle ABD is equiangular to the triangle ADC. Hence it is as BD of the triangle ABD subtending the angle BAD, is to DA subtending the angle at C of the triangle ADC, so is the same AD of the triangle ABD, subtending the angle at B to AC subtending the angle DAC of the triangle ADC, equal to that at B, and also BA subtending the right angle ADB, to AC subtending the right angle ADC; therefore the triangle ABD is similar to the triangle ADC. If, therefore, in right angled triangles, &c. Q. E. D.

Corollary.

From this it is evident, if in a right angled triangle a perpendicular be drawn from the right angle at the base, it is a mean proportional between the segments of the base; and also that each of the sides is a mean proportional between the base and its segment adjacent to that side.

Deduction.

To divide a given finite right line into two parts, such that another given right line, not greater than half of the former, shall be a mean proportional between them.

PROPOSITION IX.

PROBLEM.

From a given right line to cut off any part required.

Let AB be a given right line; it is required to cut from AB any part required.*

Let a third part be required, and from it draw any right line AC, containing any angle with AB, and take any point D in AC and make DE, EC, equal to AD, join BC, and through D draw DF parallel to BC.

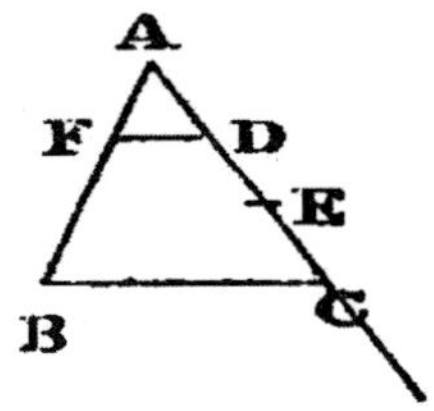

And because FD is drawn parallel to one of the sides BC of the triangle ABC; therefore, proportionally as CD is to DA so is BF to FA.[a] But CD is double of DA; therefore also BF is double of FA; hence BA is triple of AF. [a] 4. 6.

Therefore, from the given right line AB the third part required has been cut off, viz. AF. Q. E. F.

PROPOSITION X.

PROBLEM.

To divide a given right line into parts similarly to a given divided right line.

Let AB be the given divided right line, also AC cut in the points D, E, and place them so that they may contain any angle, and join CB, and through D, E, draw DF, EG, parallel to BC,[a] and through D draw DHK parallel to AB. [a] 31. 1.

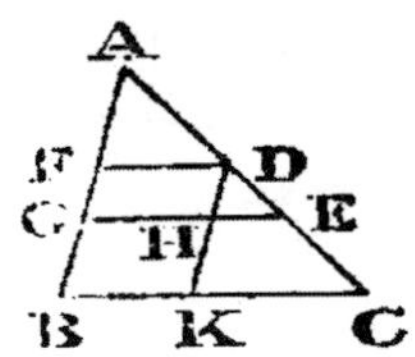

Therefore each of the figures FH, HB, is a parallelogram; hence DH is equal to FG, also HK to GB. And because HE is drawn parallel to one of the sides KC of the triangle DKC; proportionally as CE is to ED so is KH to HD; but KH is equal to BG; also HD to GF; therefore as CE is to ED so is BG to GF. Again, because FD is drawn parallel to one of the sides EG of the triangle AGE; hence proportionally it is as ED to DA so is GF to FA. But it has been de-

---

* Precisely in the same manner may any part whatever be cut off, since AF in all cases is the same part of AB which AD is of AC.

monstrated as CE is to ED so is BG to GF; therefore as CE is to ED so is BG to GF, also as ED to DA so is GF to FA. Therefore a given right line, &c. Q. E. D.

Deductions.

1. To describe a square which shall have a given ratio to a given rectilineal figure.
2. To divide a right line into three parts which shall be in harmonical progression.
3. The base, the vertical angle, and the ratio of the two sides of a triangle being given to construct it.

PROPOSITION XI.

Problem.

Two right lines being given, to find a third proportional.

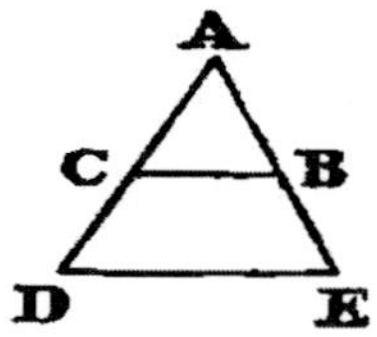

Let AB, AC, be the given right lines, and place them so that contain any angle; it is required to find Bird proportional.

For produce AB, AC, to the points D, E, and place BE equal to AC, also join BC, and
a 31. 1. through D draw DE parallel to it.[a]

Therefore BC is drawn parallel to one of the sides DE of the triangle ADE, proportionally it is as AB to
b 2. 6. [b] BE so is AC to CD. But BE is equal to AC, therefore it is as AB to AC so is AC to BE. Therefore two lines AB, AC, being given, a third proportional BE has been found. Q. E. F.

Deduction.

To determine the locus of the vertices of all the triangles, which can be described on a given base, so that each of them shall have its two sides in a given ratio.

PROPOSITION XII.

Problem.

Three given right lines being given to find a fourth proportional to them.

Let A, B, C, be three given right lines; it is required to find a fourth proportional to them.

Place the two right lines DE, DF, containing any angle EDF, and make DG equal to A, also GE equal to B, and DH equal to C; GH being joined, draw through E, EF parallel to it.[a] [a] 31. 1.

And because GH is drawn parallel to one of the sides EF of the triangle DEF, therefore as DG is to GE so is DH to HF.[b] But DG is equal to A, also GE to B, and DH to C; therefore as A is to B so is C to HF. Therefore three right lines being given A, B, C, a fourth proportional HF has been found. Q. E. F. [b] 2. 6.

Deductions.

1. Divide a given right line into two parts, so that the rectangle contained by them may be equal to a given rectangle.

2. From a given point to draw a right line to cut a given circle, so that the distances of the two intersections from the given point, shall be to each other in a given ratio.

PROPOSITION XIII.*

PROBLEM.

Two right lines being given to find a mean proportional.

Let AB, BC, be two given right lines; it is required to find a mean proportional to AB, BC.

Put them in a right line, and upon AC describe the semi-circle ADC, and draw from the point, B, BD at right angles to the right line AC,[a] and join AD, DC. [a] 11. 1.

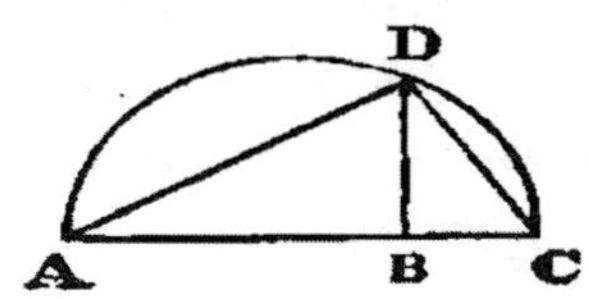

And because the angle ADC in a semi-circle is a right angle.[b] And because in the right angled triangle ADC, the perpendicular DB is drawn from the right angle to the base; DB is a mean proportional between the segments of the base.[c] Therefore the two right lines AB, BC, being given, a mean proportional BD has been found. Q. E. F. [b] 31. 3. [c] Cor. 3. 6.

---

* This is in effect the same as the last proposition of the second book.

PROPOSITION XIV.

Theorem.

If equal parallelograms have one angle of the one equal to one angle of the other, the sides about the equal angles are reciprocally proportional; and if parallelograms have one angle of the one equal to one angle of the other, and the sides about the equal angles reciprocally proportional; these parallelograms shall be equal to one another.

Let AB, BC, be equal parallelograms having the angles at B equal, and place DB, BE, in a direct line, there-
[a] 14. 1. fore FB, BG, are in a direct line;[a] then the sides about the equal angles of the parallelograms AB, BC, are reciprocally proportional, that is, as DB to BE so is GB to BF. For complete the parallelogram FE. And because the parallelogram AB is equal to the parallelogram BC, and FE is some other parallelogram; therefore
[b] 7. 5. as AB to FE so is BC to FE.[b] But as AB to FE so is DB to BE, also as BC to FE so is GB to BF; and therefore as DB to BE so is
[c] 11. 5. GB to BF.[c] Therefore the sides of the parallelograms AB, BC, are reciprocally proportional.

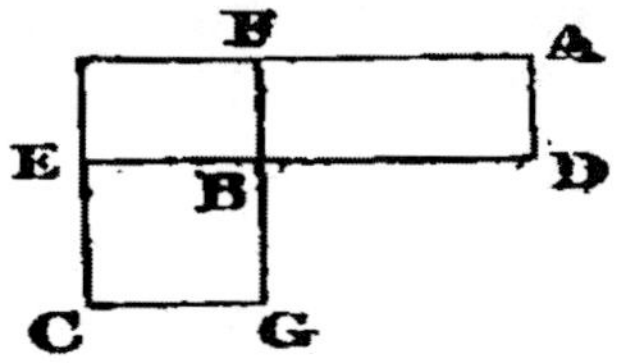

But also let the sides about the equal angles be reciprocally proportional, viz. as DB to BE so is GB to BF; then the parallelogram AB is equal to the parallelogram BC.

For because it is as DB to BE so is GB to BF, but as DB to BE so is the parallelogram AB to the parallelogram FE, also as GB to BF so is the parallelogram BC to the parallelogram FE; hence also as AB to FE so is BC to FE, therefore the parallelogram AB is equal to
[d] 9. 5. the parallelogram BC.[d] Therefore if equal parallelograms, &c. Q. E. D.

PROPOSITION XV.

Theorem.

If equal triangles have one angle of the one equal to one angle of the other, the sides about the equal angles are reciprocally proportional; and if triangles have one angle of the one equal to one angle of the other, and the sides about the equal angles reciprocally proportional, these triangles are equal.

Let ABC, ADE, be equal triangles having one angle,

BAC equal to an angle DAE; then the sides of the triangles ABC, ADE, are reciprocally proportional, that is, as CA to AD so is EA to AB. For place AC in a direct line with AD, therefore EA is in a direct line with AB.[a] And join BD. [a] 14. 1.

And because the triangle ABC is equal to the triangle ADE, but ABD is another triangle; therefore as the triangle CAB is to the triangle BAD so is the triangle ADE to the triangle BAD.[b] But as CAB to BAD so is CA to AD,[c] also as EAD is to BAD so is EA to AB; therefore the sides of the triangles ABC, ADE, are reciprocally proportional. [b] 7. 5. [c] 1. 6.

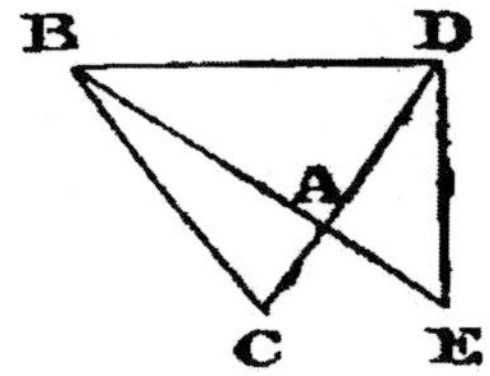

Next let the sides of the triangles ABC, ADE, be reciprocally proportional, viz. as CA to AD so is EA to AB; then the triangle ABC is equal to the triangle ADE.

For BD being joined, because it is as CA to AD so is EA to AB, but as CA to AD so is the triangle BAC to the triangle BAD, also as EA to AB so is the triangle EAD to the triangle BAD; therefore as the triangle ABC to BAD so is the triangle EAD to BAD; hence each of them ABC, ADE, has the same ratio to BAD; therefore the triangle ABC is equal to the triangle EAD. Therefore if equal triangles, &c. Q. E. D.

PROPOSITION XVI.*

THEOREM.

If four right lines be proportional, the rectangle contained under the extremes is equal to the rectangle contained under the means: and if the rectangle contained under the extremes be equal to the rectangle contained under the means, the four right lines will be proportional.

Let AB, CD, E, F, be four proportional right lines, viz. as AB to CD so is E to F; then the rectangle contained under AB, F is equal to the rectangle contained under CD, E.

For draw AG, CH, at right angles to the lines AB, CD,[a] from the points A, C, and make AG equal to F, also CH equal to E, and complete the parallelograms BG, DH. [a] 11. 1.

---

* Algebraically if A : B :: C : D, then AD = BC. For since $\frac{A}{B} = \frac{C}{D}$; AD = BC by multiplying by BD.

And because it is as AB to CD so is E to F, and E is
equal to GH, also F to AG; therefore it is as AB to CD
b 7. 5. so is CH to AG,[b] therefore the sides
about the equal angles of the pa-
rallelograms BG, DH, are recipro-
cally proportional. But the sides
about the equal angles are reci-
procally proportional of those pa-
c 14. 6. rallelograms which are equal;[c] therefore the parallelogram BG is equal to the parallelogram DH, and BG is the parallelogram contained under AB, F, for AG is equal to F, also DH is the parallelogram under CD, E, for CH is equal to E; therefore the rectangle contained under AB, F, is equal to the rectangle contained under CD, E.

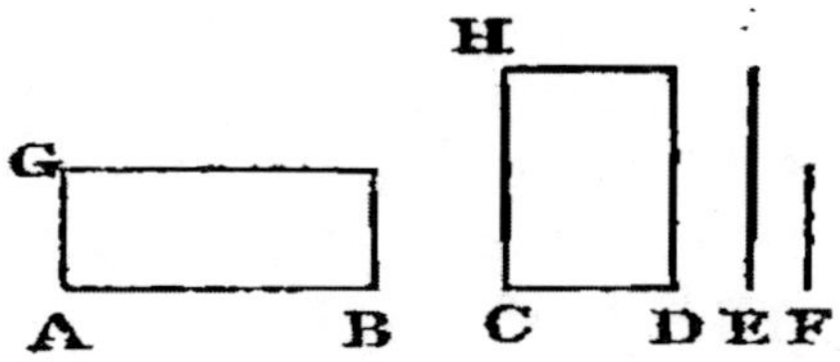

But also let the rectangle contained under AB, F, be equal to the rectangle contained under CD, E; then the four right lines shall be proportional, as AB to CD so is E to F.

For the same construction being made, because the rectangle under AB, F, equal to that under CD, E; and BG is the rectangle under AB, F, for AG is equal to F; also DH the rectangle under CD, E, since CH is equal to E, therefore BG is equal to DH, and they are equiangular. But the sides about the equal angles of equiangular parallelograms are reciprocally proportional, therefore it is as AB to CD so is CH to AG. But CH is equal to E, also AG to F; hence as AB to CD so is E to F. If therefore four right lines, &c. Q. E. D.

Deductions.

1. Of four right lines which are in continual proportion, the two extremes being given and also a line equal to the difference of the other two, to find those two lines.

2. To construct a triangle, such that the two lines including the verticle angle shall be in a given ratio, and the perpendicular from the vertex to the base equal to a given right line.

PROPOSITION XVII.*

THEOREM.

If three right lines be proportional, the rectangle contained under the extremes is equal to the square of the mean; and if the rectangle contained under the extremes be equal to the square of the mean, the three right lines will be proportional.

Let the three right lines, A, B, C, be proportional; the rectangle contained under the extremes is equal to the square of the mean.

Make D equal to B.

And because it is as A is to B so is B to C, and B is equal to D; therefore as A is to B so is D to C.[a] But if [a] 7. 5.
four right lines be proportional, the rectangle contained under the extremes is equal to the rectangle contained under the means,[b] therefore the rectangle under A, C, is equal to that under B, D. But the rectangle under B, D, is the square of B, for B is equal to D, hence the rectangle contained under A, C, is equal to the square of B. [b] 16. 6.

Next let the rectangle under A, C, be equal to the square of B; then it is as A is to B, so is B to C.

For the same construction being made, because the rectangle under A, C, is equal to the square of B, but the square of B is the rectangle under B, D, for B is equal to D; therefore the rectangle under A, C, is equal to that under B, D. But if the rectangle under the extremes be equal to the rectangle under the means, the four right lines are proportional; therefore as A is to B, so is D to C. But B is equal to D; hence as A is to B so is B to C. If therefore three right lines, &c. Q. E. D.

---

* This may also be shown algebraically, for if A : B : : B : D, then AD $= B^2$. Since $\frac{A}{B} = \frac{B}{D}$; by multiplying by BD we have AD $= B^2$. Q. E. D.

PROPOSITION XVIII.

Problem.

Upon a given right line, to describe a figure similar and similarly situated to a given right figure.

Let AB be a given right line and CE a given rectilineal figure; it is required to describe upon the right line AB a rectilineal figure similar and similarly situated to the rectilineal figure CE.

[a] 23. 1. Join DF, and upon AB make[a] the angle GAB equal to that at C, also the angle ABG equal to that at CDF; hence the remaining angle AGB is
[b] 32. 1. equal to the remaining angle CFD,[b] therefore the triangle FCD is equiangular to the triangle GAB, proportionally as FD to GB so is FC to GA and CD to AB. Again on the

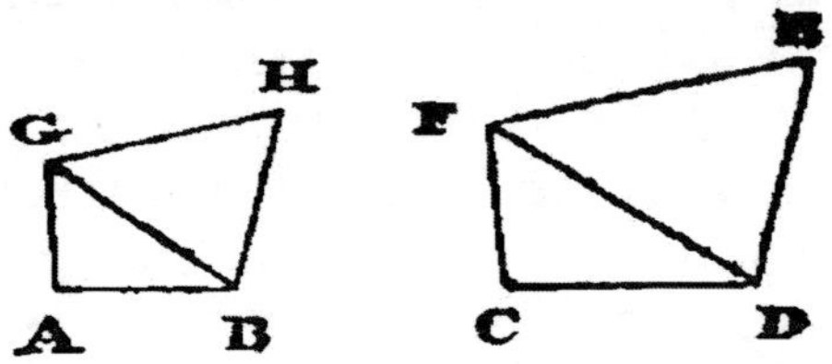

right line BG make at the points B, G, the angle BGH equal to the angle DFE, also GBH equal to FDE, hence the remaining angle at E is equal to the remaining angle at H. Therefore the triangle FDE is equiangular to the triangle GBH; proportionally as DF to GB so is FE to GH and ED to HB. But it has been shown also as FD to GB so is FC to GA and CD to AB; and therefore as FC to AG so is CD to AB and FE to GH, and consequently ED to HB. And because the angle CFD is equal to AGB, also DFE to BGH; therefore the whole CFE is equal to the whole AGH. For the same reason the angle CDE is equal to ABH. But the angle at C is equal to that at A, also the angle at E is equal to that at H. Therefore the figure AH is similar to CE and has the sides about the equal angles proportional; therefore the rectilineal figure AH is similar to the rectilineal figure CE. Therefore upon the given right line AB, the rectilineal figure AH has been described similar and similarly situated to the given rectilineal figure CE. Q. E. F.

PROPOSITION XIX.

Theorem.

Similar triangles are to one another in the duplicate ratio of their homologous sides.

Let ABC, DEF, be similar triangles having the angle at B equal to that at E, and as AB to BC so is DE to

EF, and let BC be the side homologous to EF; then the triangle ABC has a duplicate ratio to the triangle DEF, which BC has to EF.

For take[a] BG a third proportional to BC, EF, so that BC may be to EF as EF to BG, and join GA. [a] 11. 6.

And because it is as AB to BC so is DE to EF; therefore alternately it is as AB to DE so is BC to EF.[b] But as BC to EF so is EF to BG; hence also as AB to DE so is EF to BG,[c] therefore the sides about the equal angles of the triangles ABG, DEF, are reciprocally proportional. But if these triangles having one angle of the one equal to one angle of the other, and the sides about the equal angles reciprocally proportional they are equal to one another; therefore the triangle ABG is equal to the triangle DEF. And because it is as BC to EF so is EF to BG. But if three right lines be proportional, the first is said to have to the third a duplicate ratio of that which it has to the second; therefore BC has to BG a duplicate ratio of that which BC has to EF. But as BC is to BG, so is the triangle ABC to the triangle ABG; and hence the triangle ABC has to ABG a duplicate ratio of that which BC has to EF. But the triangle ABG is equal to the triangle DEF, and therefore the triangle ABC has to the triangle DEF a duplicate ratio of that which BC has to EF. Therefore similar triangles, &c. Q. E. D.

[b] 16. 5.

[c] 11. 5.

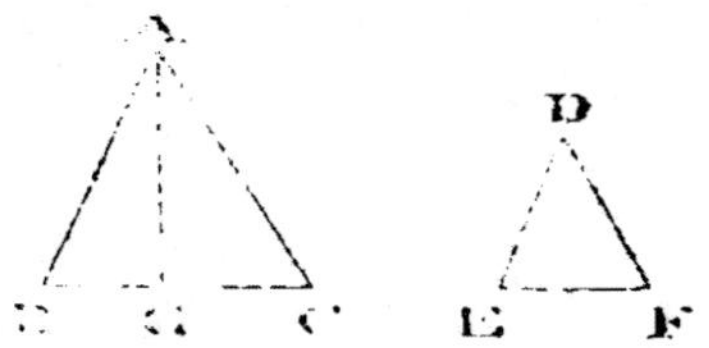

Corollary.

From this it is manifest, that if three right lines be proportional; as the first is to the third so is the triangle upon the first to the triangle similar and similarly described upon the second; because it has been shown, as CB to BG so is the triangle ABC to the triangle ABG, that is to DEF.

Deduction.

To cut off from a given triangle any part required, by a right line drawn parallel to a given right line.

PROPOSITION XX.

Theorem.

Similar polygons may be divided into similar triangles, equal in number, and homologous to the whole, and the polygons have to one another a duplicate ratio of that which their homologous sides have.

Let ABCDE, FGHKL, be similar polygons, also let AB be homologous to FG; then the polygons ABCDE, FGHKL, may be divided into similar triangles, equal in number and homologous to the wholes; also the polygon ABCDE has to the polygon FGHKL a duplicate ratio of that which AB has to FG. And because the polygon ABCDE is similar to the polygon FGHKL, the angle BAE is similar to GFL; and

a 1. 6. it is as BA to AE so is FG to FL.[a] And because there are two triangles ABE, FGL, having one angle equal to one angle, and the sides about the equal angles proportional, therefore the triangle ABE is equiangular to

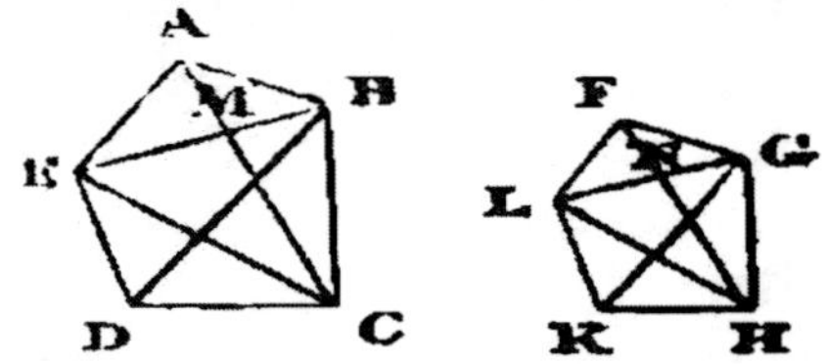

b 6. 6. the triangle FGL,[b] wherefore also it is similar;[c] hence

c 4. 6. the angle ABE is equal to FGL. But also the whole ABC is equal to the whole FGH, because they are similar polygons, therefore the remaining angle EBC is equal to the remaining angle LGH. And because the triangles ABE, FGL, are similar, it is as EB to BA so is LG to GF; but because the polygons are similar, it is as AB to BC so is FG to GH; therefore, by equality, it is as EB to BC

d 22. 5. so is LG to GH,[d] and the sides about the equal angles EBC, LGH, are proportional, therefore the triangle EBC is equiangular to the triangle LGH;[d] wherefore also the triangle EBC is similar to the triangle LGH. For the same reason the triangle ECD is similar to the triangle LHK, therefore the similar polygons ABCDE, FGHKL, contain an equal number of similar triangles.

They are also homologous to the whole, that is, the triangles are proportionals, the antecedents being ABC, EBC, ECD, and the consequents FGL, LGH, LHK, and the polygon ABCDE has a duplicate ratio to the polygon FGHKL which their homologous sides have, that is, AB to FG.

For join AC, FH.

And because the polygons are similar, the angle ABC is equal to the angle FGH, and it is as AB to BC so is

FG to GH; the triangle ABC is equiangular to the triangle GFH; hence the angle BAC is equal to GFH, also BCA to FHG. And because the angle BAM is equal to GFN, but it has been shown that ABM is equal to FGN; and therefore the remaining angle AMB is equal to the remaining angle FNG; whence the triangle ABM is equiangular to the triangle FGN. In like manner we demonstrate that the triangle BMC is equiangular to the triangle GNH; therefore proportionally as AM to MB so is FN to NG, also as BM to MC so is GN to NH; wherefore, also, by equality, as AM to MC so is the triangle ABM to MBC, and AME to EMC, for they are to one another as their bases, and therefore as one of the antecedents is to one of the consequents, so are all the antecedents to all the consequents.[e] Therefore as the triangle AMB is to BMC so is ABE to CBE. But as AMB to BMC so is AM to MC; and therefore as AM to MC so is the triangle ABE to the triangle EBC. For the same reason also as FN to NH so is the triangle FGL to the triangle GLH. And it is as AM to MC so is FN to NH; and therefore as the triangle ABE is to the triangle BEC so is the triangle FGL to the triangle GHL, and alternately as the triangle ABE is to the triangle FGL so is the triangle BEC to the triangle GFH. In like manner we show, that BD, GK being joined, as the triangle BEC to the triangle GLH so is the triangle ECD to the triangle LHK. And because it is as the triangle ABE to FGL so is EBC to FGH, also ECD to LHK; and alternately as one of the antecedents is to one of the consequents so are all the antecedents to all the consequents, therefore as the triangle ABE is to the triangle FGL so is the polygon ABCDE to the polygon FGHKL. But the triangle ABE has to the triangle FGL a duplicate ratio of that which their homologous sides have, viz. AB to FG. For similar triangles are in a duplicate ratio of their homologous sides; and therefore the polygon ABCDE has to the polygon FGHKL a duplicate ratio of that which the homologous side AB has to the homologous side FG. Therefore similar polygons, &c. Q. E. D.

[e] 12. 5.

COROLLARIES.

1. In like manner it may be demonstrated, that similar four-sided figures are to one another in the

duplicate ratio of their homologous sides; and it has been already proved in triangles. Wherefore universally similar rectilineal figures are to one another in the duplicate ratio of their homologous sides. Q. E. F.

2. And if to AB, FG, we take a third proportional X; AB has to X a duplicate ratio of that which AB has to FG. But one four-sided figure or polygon has to another four-sided figure or polygon, a duplicate ratio of that which their homologous sides have, that is AB to FG, which was also proved in triangles. Therefore universally it is manifest, if three right lines be proportional, as the first is to the third, so is any rectilineal figure upon the first to the similar and similarly described rectilineal figure upon the second.

Deduction.

Any regular polygon inscribed in a circle, is a mean proportional between the inscribed and circumscribed regular polygon of half the number of sides.

PROPOSITION XXI.

Theorem.

Rectilineal figures that are similar to the same rectilineal figure, are also similar to one another.

For let each of the rectilineal figures A, B, be similar to C, then A, B, are similar to one another.

For because A is similar to C, they are equiangular,
a 1 Def. 6. and have the sides about the equal angles proportional.[a]

Again, because B is similar to C, they are also equiangular, and have the sides about the equal angles proportional, therefore each

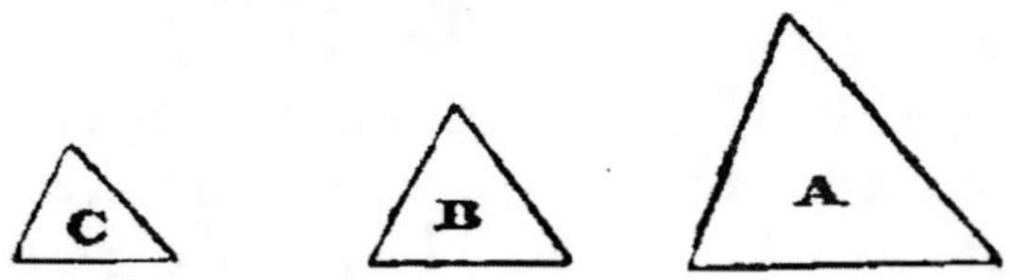

of them A, B, is equiangular to C, and has the sides about the equal angles proportional, therefore A is similar to B. Therefore rectilineal figures, &c. Q. E. D.

PROPOSITION XXII.

THEOREM.

If four right lines be proportional, the similar rectilineal figures, and similarly described upon them, shall also be proportional, and if the similar rectilineal figures, similarly described upon four right lines, be proportional, the right lines themselves shall be proportional.

Let the four right lines AB, CD, EF, GH, be proportional, as AB to CD so is EF to GH, and upon AB, CD, describe the similar and similarly placed rectilineal figures KAB, LCD, also upon EF, GH, the similar and similarly placed rectilineal figures MF, NH; then as KAB is to LCD so is MF to NH.

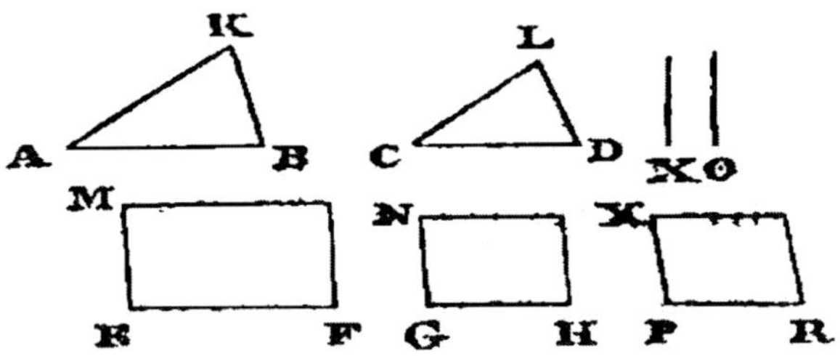

For take X a third pro-
portional to AB, CD,[a] also [a] 11. 6.
O a third proportional to
EF, GH. And because it
is as AB to CD so is EF to
GH, also as CD is to X so
is GH to O,[b] therefore by [b] 11. 5.
equality it is as AB to X so
is EF to O.[c] But as AB to X so is KAB to LCD, also as [c] 22. 5
EF to O so is MF to NH;[d] and therefore as KAB to [d] 2 Cor. 20. 6.
LCD so is MF to NH. But also let it be as KAB to LCD
so is MF to NH; then also as AB to CD so is EF to GH.

For make[e] as AB to CD so is EF to PR, and KAB, [e] 12. 6.
LCD, are described upon AB, CD, similar and similarly placed, also upon EF, PR, the figures MF, XR, similar and similarly situated; therefore it is as KAB to LCD so is MF to XR. But, by construction, as KAB to LCD so is MF to NH; and therefore also as MF to XR so is MF to NH; therefore MF has the same ratio to each of them NH, XR; therefore NH is equal to XR. But it is similar and similarly placed; hence GH is equal to PR. And because it is as AB to CD so is EF to PR, but PR is equal to GH; therefore it is as AB to CD so is EF to GH. If therefore four right lines, &c. Q. E. D.

Deduction.

If any two chords of a circle intersect each other, the right lines joining their extremities shall cut off equal segments from the chord which passes through the common intersection of the two former chords, and is there bisected.

PROPOSITION XXIII.

Theorem.

Equiangular parallelograms are to one another in a ratio compounded of the ratios of their sides.

Let AC, CF, be equiangular parallelograms having the angle BCD equal to ECG, then the parallelogram AC is to the parallelogram CF in a ratio compounded of the ratios of their sides, of that which BC has to CG and of that which DC has to CE.

For place BC in a direct line with CG; therefore also DC is in
a 14. 1. a direct line with CE,[a] and complete the parallelogram DG, let K be a certain right line, and make as BC is to CG so is K to L, also

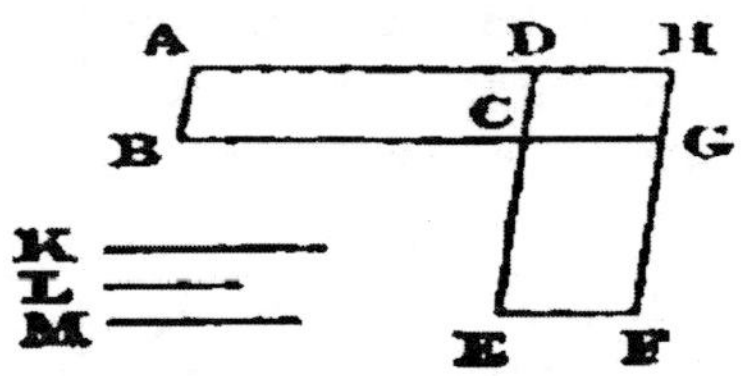

b 12. 6. as DC is to CE so is L to M.[b] Therefore the ratios of K to L and of L to M are the same as the ratios of the sides both of BC to CG and of DC to CE. But the ratio of K to M is composed of the ratio of K to L and of the ratio of L to M; wherefore K has to M the ratio compounded of their sides. And because it is as BC to
c 1. 6. CG so is the parallelogram AC to CH.[c] But as BC to CG so is K to L; and therefore as K to L so is AC to CH. Again, because it is as DC to CE so is the parallelogram CH to CF; but as DC to CE so is L to M, and hence as L is to M so is the parallelogram CH to the parallelogram CF.

Therefore because it has been shown as K to L so is the parallelogram AC to the parallelogram CH, also as L to M so is the parallelogram CH to the parallelogram CF, therefore, by equality, it is as K to M so is the
d 22. 5. parallelogram AC to the parallelogram CF.[d] But also K has to M the ratio compounded of the ratios of their sides; hence also AC is to CF in the ratio compounded of the ratio of their sides. Therefore equiangular parallelograms, &c. Q. E. D.

PROPOSITION XXIV.*

THEOREM.

Parallelograms which are about the diameter of any parallelogram are similar to the whole and to one another.

Let ABCD be a parallelogram, and AC its diameter, and let FG, HK, be parallelograms about AC; then each of the parallelograms FG, HK, is similar to the whole ABCD, and to one another.

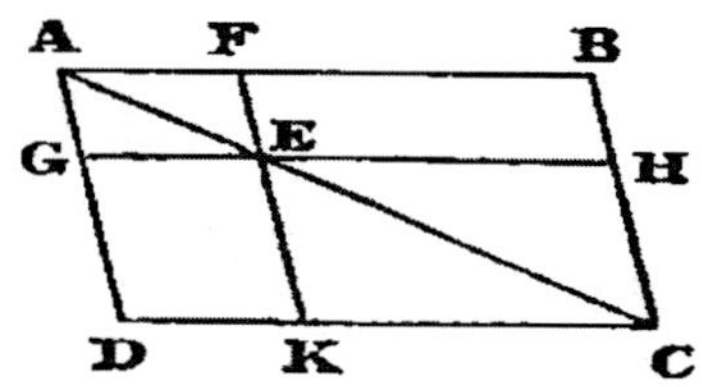

For because EF is drawn parallel to one of the sides BC of the triangle ABC, proportionally, it is as BF to FA so is CE to EA. Again, because EG is drawn parallel to one of the sides CD of the triangle ACD, proportionally, it is as CE to EA so is DG to GA. But as CE to EA so is BF to FA; and therefore as BF to FA so is DG to GA, and by composition, as BA to AF so is DA to AG, and alternately as BA to AD so is FA to AG; therefore the sides of the parallelograms ABCD, FG, are proportional about the common angle BAD. And because GE is parallel to DC, the angle AGE is equal to ADC, also GEA to DCA, and DAC common to the two triangles ADC, AGE; therefore the triangle ADC is equiangular to the triangle AGE. For the same reason, the triangle ACB is equiangular to the triangle AFE; and therefore the whole parallelogram ABCD is equiangular to the parallelogram FG; hence proportionally it is as AD to DC so is AG to GE. But as DC to CA so is GE to EA, also as AC to CB so is AE to FE, and also as CB to BA so is FE to FA; and because it has been shown as DC to CA so is GE to EA, also as AC to CB so is AE to FE; therefore, by

* From hence it is observable, that the parallelograms about the diameter are like figures having their sides to one another directly proportional, and the complements are equal parallelograms, having their sides reciprocally proportional to one another. Prop. 43, lib. 1. Prop. 14, lib. 6. Moreover each of the complements is a mean proportional between the parallelograms about the diameter, by prop. 1, lib. 6, and prop. 38, lib. 1, which also are to one another in a duplicate ratio of their homologous sides. Prop. 20, lib. 6. Pelitarius has very well spoken upon the excellency of this proposition. Hanc ego figuram soleo vocare mysticam. Ex ea enim, velut eo locupletissimo promptuario, innumerabiles exeunt demonstrationes, quod cum magna voluptate prospiciet, qui in re Geometrica serio se exercebit.

equality, it is as DC to BC so is GE to FE. Therefore the sides about the equal angles of the parallelograms ABCD, FG, are proportional; hence the parallelogram ABCD is similar to the parallelogram FG. For the same reason, also the parallelogram ABCD is similar to the parallelogram HK; therefore each of the parallelograms FG, HK, is similar to the parallelogram ABCD. But similar rectilineal figures, which are the same to the same rectilineal figure are the same to one another; and hence the parallelogram FG is similar to the parallelogram HK; wherefore parallelograms which are, &c. Q. E. F.

PROPOSITION XXV.

Problem.

To describe a rectilineal figure similar to one and equal to another given rectilineal figure.

Let ABC be the given rectilineal figure to which the figure to be described is required to be similar, also D that to which it is required to be equal; it is, therefore, required to describe a figure similar to ABC, and also equal to D.

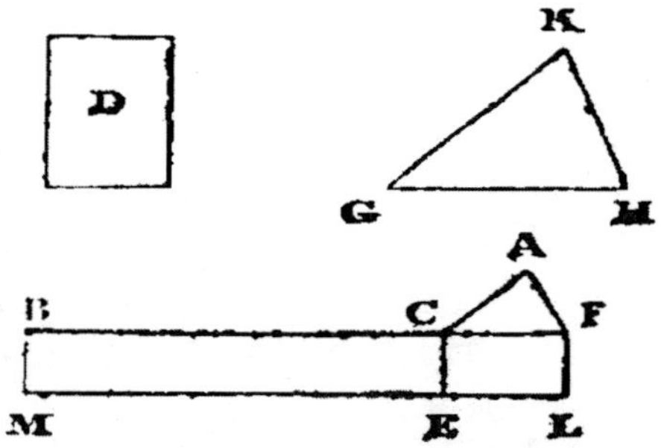

For to BC apply the parallelogram BE equal to the triangle ABC, also to CE the parallelogram CM equal to D, containing an angle ECF, which is equal to CBL; therefore BC is in a direct
a 14. 1. line with CF,[a] also LE to EM. And
b 13. 6. find[b] GH a mean proportional between BC, CF, and describe upon GH the figure KGH similar and similarly situated to
c 18. 6. ABC.[c]

And because it is as BC to GH so is GH to CF, but if three right lines be proportional, as the first is to the third, so is the figure described upon the first to the similar and similarly described figure upon the second; therefore it is as BC to CF so is the triangle ABC to the triangle KGH. But also as BC to CF so is the parallelo-
d 20. 6. gram BE to the parallelogram EF;[d] therefore alternately as the triangle ABC is to the parallelogram BE so is the triangle KGH to the parallelogram EF. But the triangle ABC is equal to the parallelogram BE, and

therefore the triangle KGH is equal to the parallelogram EF. But the parallelogram EF is equal to D, and hence KGH is equal to D. And KGH is similar to ABC. Therefore a rectilineal figure KGH is described similar to the given rectilineal figure ABC, and equal to another given rectilineal figure D. Q. E. F.

PROPOSITION XXVI.

THEOREM.

If from a parallelogram, a parallelogram be taken away similar and similarly situated to the whole, and both having one common angle; they will also have the same common diameter.

For let the parallelogram AEFG be taken from the parallelogram ABCD, similar and similarly situated to it, both having the common angle DAB; then both the parallelograms ABCD, AEFG, are about the same diameter.

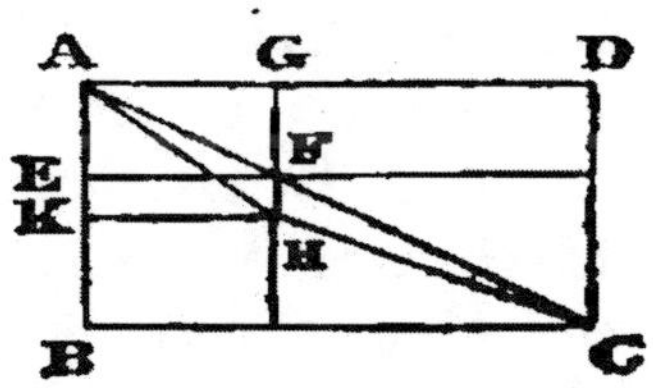

For if not, let, if possible, AHC be the diameter of ABCD, and produce GF to H, and through H draw HK parallel to either of them AD or BC.

Therefore because the parallelograms ABCD, KG, are about the same diameter, ABCD is similar to KG;[a] therefore as DA is to AB so is GA to AK. But because the parallelograms ABCD, EG, are similar to one another, it is as DA to AB so is GA to AE; and therefore as GA to AK so is GA to AE; hence GA has the same ratio to each of them AK, AE; therefore AE is equal to AK, the less to the greater, which is impossible; hence the parallelograms ABCD, KG, are not about the same diameter; wherefore ABCD, AEFG, are about the same diameter. If, therefore, from a parallelogram, &c. Q. E. D.

[a] 24. 6.

PROPOSITION XXVII.

Theorem.

Of all parallelograms applied to the same right line, and deficient by parallelograms similar and similarly situated to that described upon half the line, that which is applied to the half and similar to its defect is the greatest.

Let AB be a right line, and bisect it in C, and apply to the same right line AB, the parallelogram AD, deficient by the parallelogram CE, similar and similarly situated to that described upon the half line AB, that is the parallelogram described upon CB; then of all the parallelograms applied to AB, and deficient by parallelograms similar and similarly situated to CE, the greatest is AD. For apply the parallelogram AF to the right line AB deficient by the parallelogram KH, similar and similarly situated to CE; AD is greater than AF.

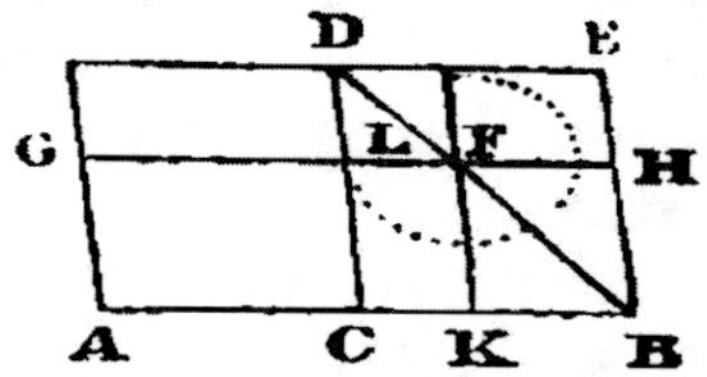

For because the parallelogram CE is similar to the parallelogram KH,
[a] 26. 6. they are about the same diameter,[a] draw their diameter DB and describe the figure. Therefore because CF is
[b] 43. 1. equal to FE,[b] add KH which is common; hence the whole CH is equal to the whole KE. But CH is equal to CG, since AC is equal to CB, and GC is, therefore, equal to EK. Add CF, which is common; therefore the whole AF is equal to the gnomon LMN; wherefore also the parallelogram CE, that is AD, is greater than the parallelogram AF.

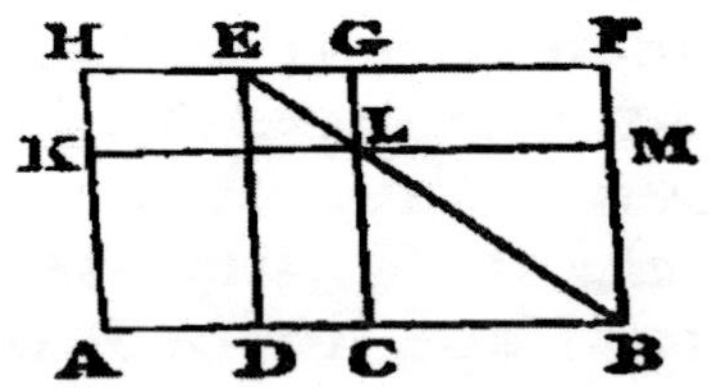

For, again, let AB be bisected in C, and the parallelogram AL applied, deficient by the figure CM, and again apply to AB the parallelogram AE, deficient by DF, similar and similarly situated to CM described upon half of the line AB, then is the parallelogram AB applied to half the line greater than AE. For because DF is similar to CM, they are about the same diameter; let EB be their diameter, and describe the figure.

And because LF is equal to LH, for FG is equal to GH; hence LF is greater than KE. But LF is equal to DL; therefore also DL is greater than EK. Add KD, which is common; therefore the whole AL is greater than the whole AE. Whence of all parallelograms, &c. Q. E. D.

PROPOSITION XXVIII.

PROBLEM.

To apply a parallelogram to a given right line equal to a given rectilineal figure, being deficient by a parallelogram similar to a given parallelogram. But the given right lined figure to which the parallelogram is to be applied equal, must not be greater than that applied parallelogram, which is described upon half the line; its defect being similar to the defect of that which is to be applied; that is, to the given parallelogram.

Let AB be a given right line, also C a given rectilineal figure, to which the parallelogram applied to AB is required to be equal, not greater than that applied to half the line, the parallelograms being the deficiencies of these parallelograms being similar. It is required to apply to the given right line AB a parallelogram equal to the given rectilineal figure C, deficient by a parallelogram similar to D.

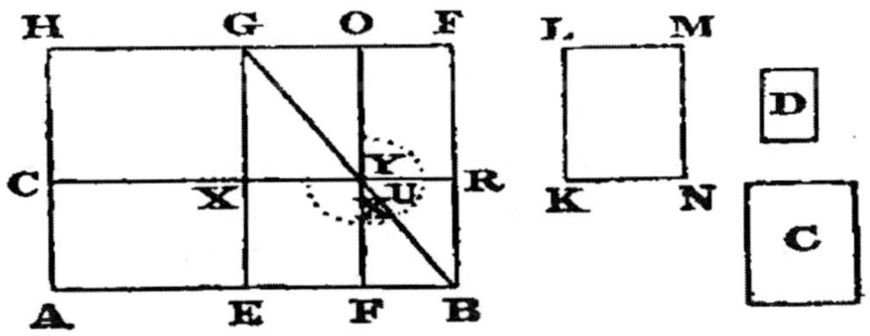

Bisect AB in the point E, and describe upon EB the parallelogram EBFG, similar and similarly situated to D,[a] and complete the parallelogram AG, AG is either equal to C or greater than it, by reason of the limitation. And if AG is equal to C, the thing proposed will be done, for to the given right line AB the parallelogram AG will be applied equal to the given rectilineal figure C, and deficient by the parallelogram EF similar to D. But if not, HE is greater than C. Wherefore HB is greater than C, make a parallelogram KLMN equal to the excess of HE, similar and similarly situated to D.[b] But D is similar to GB; and therefore KM is similar to GB. Let thence KL be homologous to GE, also LM to GF. And because GB is equal to C, KM, therefore GB is greater than KM; hence also GE is greater than LK, also GF than LM. But GX equal to KL, also GH equal to LM, and complete the parallelogram XGOP, therefore GP is equal and similar to KM. But KM is similar to GB, and GP is therefore similar to GB; hence GP and CP are

[a] 18. 6.

[b] 25. 6.

about the same diameter. Let GPB be their diameter, and describe the figure.

And because BG is equal to CKM, of which GP is equal to KM; hence the remaining gnomon UYX is equal to C. And because OR is equal to XS, add PB which is common; therefore the whole OB is equal to the whole XB. But XB is equal to TE. And because the side AE is equal to the side EB; and TE is therefore equal to OB. Add XS, which is common, therefore the whole TS is equal to the gnomon UYX. But UYX has been shown to be equal to C, and therefore AP is equal to C.

Therefore to the given right line AB the parallelogram SU has been applied equal to the given rectilineal figure C, deficient by the parallelogram PB similar to D, whence PB is similar to GP. Q. E. F.

PROPOSITION XXIX.

Problem.

To a given right line to apply a parallelogram equal to a given rectilineal figure, exceeding by a parallelogram similar to a given parallelogram.

Let AB be the given right line, also C the rectilineal figure to which the parallelogram applied to AB is required to be equal, and D that to which it is to be similar; it is required, therefore, to the right line AB to apply a parallelogram equal to the rectilineal figure C, exceeding by a parallelogram similar to D.

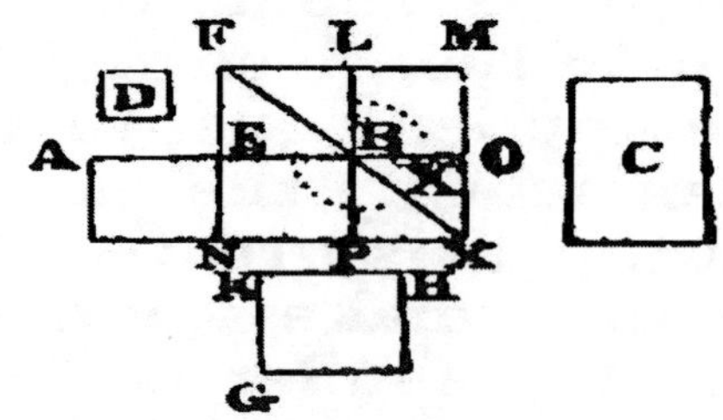

[a] 18. 6. Bisect AB in E, and describe upon EB, the parallelogram BF similar and similarly situated to D,[a] and BF equal to C, also describe GH similar and similarly situated to D; therefore GH is similar to EL. But let KH be homologous to FL, also KG to FE. And because GH is greater than FB, therefore KH is also greater than FL, also KG than FE. Produce FL, FE, and let FLM be equal to KH, also FEN to KG, and complete MN; therefore MN is equal and similar to GH. But GH is similar to EL, and hence MN is similar to EL; and EL, MN, are about the same diameter. Draw their diameter FX, and describe the figure.

And because GH is equal to EL, C, but GH is equal to MN, and MN is therefore equal to EL, C. Take away EL, which is common. Therefore the remaining gnomon VQY is equal to C. And because AE is equal to EB, AN is also equal to NB, that is to LO. Add EX, which is common; therefore the whole AX is equal to the gnomon VXY. But the gnomon VXY is equal to C; and AX is therefore equal to C. Therefore to the given right line AB, a parallelogram C has been applied, exceeding by the parallelogram PO similar to D, because OP is similar to D.[b] Q. E. F. [b] 24. 6.

PROPOSITION XXX.

PROBLEM.

To cut a given finite right line in extreme and mean ratio.

Let AB be a given finite right line; it is required to cut the given right line AB in extreme and mean ratio. Describe upon AB the square BC,[a] and apply to AC the parallelogram CD equal to BC, exceeding by the figure AD similar to BC.[b] [a] 46. 1. [b] 29. 6.

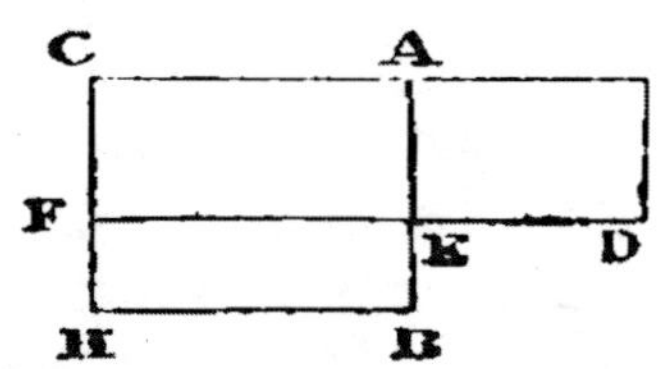

But BC is a square, hence AD is also a square. And because BC is equal to CD, take away CE, which is common; therefore the remainder BF is equal to the remainder AD, but it is also equiangular to it, therefore the sides of them BF, AD about the equal angles are reciprocally proportional; hence it is as FE to ED so is AE to EB. But FE is equal to AC, that is, to AB, also ED to AE; therefore it is as BA to AE so is AE to EB. But AB is greater than AE; hence also AE is greater than EB. Therefore the right line AB has been cut in E in extreme and mean ratio, and AE is its greater segment. Q. E. D.

Otherwise.

Let AB be a given right line, it is required to cut AB in extreme and mean ratio.

Divide AB in C, so that the rectangle under AB, BC, may be equal to the square of AC.[c] [c] 11. 2.

And because the rectangle under AB, BC, is equal to the square of CA; hence

[d] 17. 6. it is as AB to AC so is AC to CB.[d] Therefore AB has been cut in C in extreme and mean ratio. Q. E. D.

Deductions.

1. A given right line being cut in extreme and mean ratio, if from the greater segment the less be taken, the greater segment also will thus be cut in extreme and mean ratio; and if a right line, equal to the greater segment, be added to the given line, the line, which is made up of the given line and this segment, is also cut in extreme and mean ratio.

2. Upon a given right line as an hypothenuse to describe a right-angled triangle, which shall have its three sides continual proportionals.

PROPOSITION XXXI.

Theorem.

In right-angled triangles, the figure described upon the side subtending the right angle is equal to the figures, similar and similarly described upon the sides, containing the right angle.

Let ABC be a right-angled triangle, having the right angle BAC; then the figure described upon BC is equal to the similar and similarly described figures upon BA, AC. Draw the perpendicular AD.

And because in the right-angled triangle BAC, a perpendicular AD is drawn upon the base BC from the right angle at A; therefore the triangles ABD, ADC, are similar to the whole ABC, and to one another. And because ABC is similar to ABD, therefore

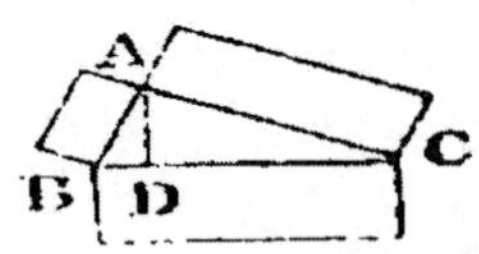

[a] 4. 6. it is as CB to BA so is BA to BD.[a] And because there are three proportionals, it is as the first to the third so is the figure described upon the first to the similar and similarly described figure upon the second; hence as CB to BD so is the figure upon CB to the similar and similarly described figure upon BA. For the same reason, as BC is to CD so is the figure upon BC to the similar and similarly described figure upon CA; wherefore also as BC is to BD, DC, so is the figure upon BC to those

upon CA, BA. But BC is equal to BD, DC; therefore also the figure upon BC is equal to the similar and similarly described figures upon BA, AC. Therefore in right-angled triangles, &c. Q. E. D.

PROPOSITION XXXII.

THEOREM.

If two triangles having two sides of the one proportional to two sides of the other, be joined at one angle, so that their homologous sides be parallel; the remaining sides of those triangles shall be in one right line.

Let ABC, CDE, be two triangles, having the two sides BA, AC, proportional to the two sides CD, DE, viz. as AB to AC so is DC to DE, also AB parallel to DC, and AC to DE; then BC, CE, are in one right line.

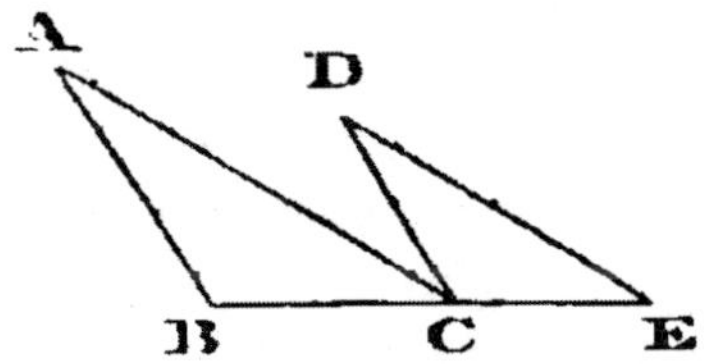

For because AB is parallel to DC, and AC falls upon them, the alternate angles, BAC, ACD, are equal to one another.[a] For the same reason, also, CDE is equal to ACD; wherefore also BAC is equal to CDE. And because there are two triangles ABC, DCE, having the angle at A equal to the angle at D, and the sides about the equal angles proportionals, viz. as BA to AC so is CD to DE. Therefore the triangle ABC is equiangular to the triangle CDE;[b] hence the angle ABC is equal to DCE. But it has been shown that ACD is equal to BAC: hence the whole ACE is equal to the two ABC, BAC; add the common angle ACB; therefore the angles BAC, ABC, BCA, are equal to two right angles,[c] and ACE, ACB. To any right line AC, and at any point C, two right lines BC, CE, not placed at the same parts, make the adjacent angles ACE, ACB, equal to two right angles; hence BC is in the same right line with CE.[d]

[a] 29. 1.

[b] 6. 6.

[c] 32. 1.

[d] 14. 1.

PROPOSITION XXXIII.

Theorem.

In equal circles, angles have the same ratio as the circumferences on which they stand, whether they be at the centres, or at the circumferences; and so also are the sectors, as being at the centres.

Let ABC, DEF, be equal circles, and BGC, EHF, at their centres G, H, also BAC, EDF, at the circumferences; then as the circumference BC is to the circumference EF so is the angle BGC to EHF, and BAC to EDF, also the sector GBC to the sector HEF.

Take any number of circumferences CK, KL, each equal to BC, and any number whatever FM, MN, each equal to EF, and join GK, GL, HM, HN. And because the circumferences BC, CK, KL, are equal to one another, and the angles BGC, CGK, KGL,

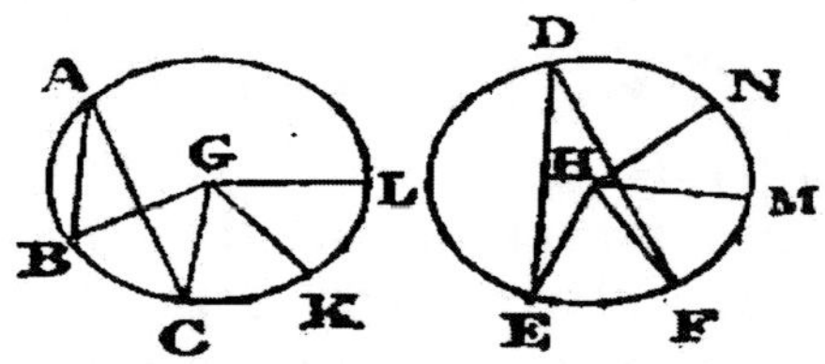

[a] 27. 3. equal to one another.[a] Whatsoever multiple the circumference BL is of BC, the same multiple is the angle BGL of BGC. For the same reason, whatsoever multiple the circumference EN is of EF, the same multiply is the angle EHN of EHF. If therefore the circumference BL be equal to the circumference EN, the angle BGL is also equal to EHN;[b] and

[b] 27. 3.

if the circumference BL be greater than the circumference EN, the angle BGL is greater than the angle EHN; and if less, less; hence there being four magnitudes, the two circumferences BC, EF, also the two angles BGC, EHF, of the circumference BC and of the angle BGC are taken any equimultiples whatsoever, viz. the circumference BL and the angle, BGL also of the circumference EF and angle EHF, viz. the circumference EN and the angle EHN; and it has been shown if the circumference BL exceed the circumference EN, the angle BGL will exceed the angle EHN; if equal, equal; and if less, less; therefore it is as the circumference BC to EF so is the angle BGC to EHF.[c] But as the angle BGC to EHF so is BAC to EDF, for each is double of each; and hence as the circumference BC is to the circumference EF so is the angle BGC to EHF and BAC to EDF. Therefore in equal circles, angles have the same ratio as the circumferences on which they stand, whether at the centres or at the circumferences. Q. E. D.

[c] 5 Def. 5.

Again, as the circumference BC is to the circumference EF so is the sector G B C to the sector HEF.

For join BC, CK, and take any points X, O, in the circumference, BC, CK, and join BX, XC, CO, OK. And because the two BG, GC, are equal to the two CG GK, and they comprehend equal angles, the base BC is also equal to CK; therefore the triangle BGC is equal to the triangle GCK. And because the circumference BC is equal to the circumference CK, also the remaining circumference of the whole circle is equal to the remaining circumference of the whole circle; wherefore also the angle BXC is equal to the angle COK; hence the segment BXC is similar to the segment COK; and they are upon equal right lines BC, CK. But similar segments of circles upon equal right lines are equal to one another; hence the segment BXC is equal to the segment COK. But the triangle BGC is equal to the triangle GCK; and therefore the whole sector GBC is equal to the whole sector GCK. For the same reason the sector GKL is equal to each of them GKC, GCB; hence the three sectors GBC, GCK, GKL, are equal to one another. For the same reason also the sectors HEF, HFM, HMN, are equal to one another; therefore whatsoever multiple the circumference BL is of the circumference BC, the same multiple is the sector GBL of the sector GBC. For the same reason also whatsoever multiple the circumference EN is of the circumference EF, the same multiple is the sector HEN of the sector HEF; if therefore the circumference BL is equal to the circumference EN, the sector BGL is also equal to the sector HEN; and if the circumference BL exceed the circumference EN, the sector GBL will also exceed the sector HEN; and if less, less. Hence there being four magnitudes, the two circumferences BC, EF, also the two sectors GBC, HEF, and any equimultiples of the circumference BC and of the sector GBC, are taken, viz. the circumference BL and the sector GBL; also any equimultiples of the circumference EF and sector HEF, viz. the circumference EN, and sector HEN. And it has been shown that if the circumference BL exceed the circumference EN, the

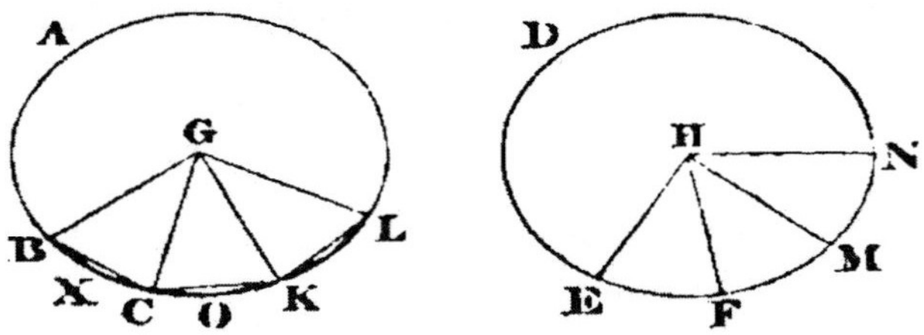

sector GBL will also exceed the sector HEN; if equal, equal; and if less, less; therefore as the circumference BC is to EF so is the sector GBC to the sector HEF. Therefore in equal circles, &c. Q. E. D.*

Deduction.

To trisect a given circle, by dividing it into three equal sections.

---

* The latter part of this proposition was added by Theon, as he informs us in his Commentaries on Ptolemy's Almagest. He says, *ἔτι δὲ καὶ οι τομεῖς· and moreover also the sectors;* which are the only words of Theon added to Euclid's proposition, for what is subjoined *ατεπροστοῖς ηεντροῖς συνιστaμενοι,* when *constituted at the centres,* must be some marginal note very absurdly put in, as supposing there were another kind of sectors, besides what are stated at the centre of the circle, according to def. 10th, lib. 3d. Indeed the figures at the circumference are not, as their angles are, in the same ratio with the arcs on which they insist; but these figures are not called *sectors*, neither have they any note or name in Geometry to give occasion for such a needless caution—an oversight too great for Theon to be guilty of.

FINIS.